International Conference on Semiconductor Nanostructures for Optoelectronics and Biosensors (IC SeNOB 2016)

EPJ Web of Conferences Volume 133 (2017)

Rzeszow, Poland
22 - 25 May 2016

Editors:

Malgorzata Pociask-Bialy
Michal Marchewka
Dariusz Ploch
Eugen Sheregii

ISBN: 978-1-5108-3513-9

Printed from e-media with permission by:

Curran Associates, Inc.
57 Morehouse Lane
Red Hook, NY 12571

Some format issues inherent in the e-media version may also appear in this print version.

Printed by Curran Associates, Inc. (2017)

For additional information, please contact EDP Sciences – Web of Conferences
at the address below.

EDP Sciences – Web of Conferences
17, Avenue du Hoggar
Parc d'Activité de Courtabœuf
BP 112
F-91944 Les Ulis Cedex A
France

Phone: +33 (0) 1 69 18 75 75
Fax: +33 (0) 1 69 28 84 91

contact-edps@webofconferences.org

Additional copies of this publication are available from:

Curran Associates, Inc.
57 Morehouse Lane
Red Hook, NY 12571 USA
Phone: 845-758-0400
Fax: 845-758-2633
Email: curran@proceedings.com
Web: www.proceedings.com

TABLE OF CONTENTS

INTERNATIONAL ADVISORY COMMITTEE

Jean-François Berret, FR
Robert Cohn, USA
Tomasz Dietl, PL
Danuta Leszczyńska, USA
Marian Grynberg, PL,
Rolf Hempelmann, GE
Stefan Jurga, PL
Volodymyr Litovchenko, UA
Andrzej Kisiel, PL
Jacek Kossut, PL
Henri Mariette, FR
Robert Parfieniev, RU
Leszek Sirko, PL
Valentyn Smyntyna, UA
Marek Szymoński, PL
Mykola Tkach, UA
Robert Triboulet, FR
Evgeny Sherman, ES
Włodzimierz Zawadzki, PL
Paweł Zięba, PL

PROGRAM COMMITTEE

Józef Barnaś, PL
Charles Becker, GE
Mikołaj Berchenko, PL
Alberta Bonanni, AT
Maciej Bugajski, PL
Ryszard Buczko, PL
Mykhailo Gonchar, UA
Ihor Izhnin, UA
Jacek Jasiński, USA
Grzegorz Karczewski, PL
Takaaki Koga, JP
Yaroslav Lepikh, UA
Roman Major, PL
Augusto Marcelli, IT
Jan Misiewicz, PL
Teresa Paronyan, USA
Małgorzata Pociask-Biały, PL
Roman Sobolewski, USA
Eugeniusz Sheregii, PL
Frank Szmulowicz, USA
Igor Tralle, PL
Jerzy Wróbel, PL
Tomasz Wojtowicz, PL
Zbigniew Żytkiewicz, PL

LOCAL COMMITTEE

Co-Chairs:
Eugeniusz Szeregij
Małgorzata Pociask-Biały

Secretary:
Dariusz Płoch

Iryna Berezovska
Ewa Bobko
Przemysław Cypryś
Jakub Grendysa
Michał Marchewka
Danuta Piotrowska
Paweł Śliż
Małgorzata Trzyna
Grzegorz Tomaka
Marek Witalec
Renata Wojanowska-Nowak
Dariusz Żak

Statement of Peer review

In submitting conference proceedings to EPJ Web of Conferences, I certify to the Publisher that I adhere to the **Policy on Publishing Integrity** of the journal in order to safeguard good scientific practice in publishing.

1. All articles have been subjected to peer review administered by the proceedings editors.
2. Reviews have been conducted by expert referees, who have been requested to provide unbiased and constructive comments aimed, whenever possible, at improving the work.
3. Proceedings editors have taken all reasonable steps to ensure the quality of the materials they publish and their decision to accept or reject a paper for publication has been based only on the merits of the work and the relevance to the journal.

Title, date and place of the conference

International Conference on Semiconductor Nanostructures for Optoelectronics and Biosensors ICSeNOB2016, 22-25 May 2016 Rzeszow, Poland

Proceedings editor(s):

Malgorzata Pociask – Bialy
Michal Marchewka
Dariusz Ploch
Eugen Sheregii

Date and editor's signature

The IC SeNOB 2016 was held in Rzeszow, Poland on May 22-25, 2016.

Participants' contributions were presented and discussed over 7 concurrent sessions:

A – MBE growth of II-VI and III-V nanostructures (QW, MQW, QWires) for optoelectronics and biosensors
B - Characterisation of II-VI and III-V nanostructures (optical properties and electron transport);
C – Applications for optoelectronics (photovoltaics, IR- and Thz- detectors, lasers);
D – Topological Insulators;
E – Applications for quantum logics;
F – Biosensors based on II-VI and III-V nanostructures;
G – Numeric methods and simulation for characterization of II-VI and III-V nanostructures (optical properties and electron transport),

The conference was attended by 113 registered participants (16 scientific sessions, 47 speakers, 38 posters; 14 plenary speakers, 18 invited speakers and 17 speakers presented the conference scientific papers) from 12 countries in Europe, America and Asia as well as 34 guests from the authorities of the University of Rzeszow and representatives of regional authorities (by Marshal's Office, deputy of the Polish Parliament), and representatives of cooperating with the University businesses and companies representing the airline industry, colleagues from the Faculty of Mathematics and Natural Sciences and the Faculty of Biology and Agriculture, and others.

There were presented the results of the achievements of scientific research conducted by the research teams. More information about the program and conference speakers you can find on the website: http://nano.rzeszow.pl/general_information.php

Properties of arsenic–implanted $Hg_{1-x}Cd_xTe$ MBE films

Igor I. Izhnin[1,2], *Alexandr V.* Voitsekhovskii[2], *Alexandr G.* Korotaev[2], *Olena I.* Fitsych[3], *Oleksandr Yu.* Bonchyk[4], *Hrygory V.* Savytskyy[4], *Karim D.* Mynbaev[5,6,*], *Vasilii S.* Varavin[7], *Sergey A.* Dvoretsky[7], *Maxim V.* Yakushev[7], *Rafal* Jakiela[8], and *Malgorzata* Trzyna[9]

[1]Scientific Research Company "Carat", 79031 Lviv, Ukraine
[2]National Research Tomsk State University, 634050 Tomsk, Russia
[3]Hetman Petro Sahaidachny National Army Academy, 79012 Lviv, Ukraine
[4]Ya.S. Pidstryhach Institute for Applied Problems of Mechanics and Mathematics NASU, 79060 Lviv, Ukraine
[5]Ioffe Institute, 194021 Saint-Petersburg, Russia
[6]ITMO University, 197101 Saint-Petersburg, Russia
[7]Rzhanov Institute of Semiconductor Physics, 630090 Novosibirsk, Russia
[8]Institute of Physics, Polish Academy of Sciences, 02-668 Warsaw, Poland
[9]Faculty of Mathematics and Natural Sciences, University of Rzeszow, 35-310 Rzeszow, Poland

Abstract. Defect structure of arsenic-implanted $Hg_{1-x}Cd_xTe$ films (x=0.23–0.30) grown with molecular-beam epitaxy on Si substrates was investigated with the use of optical methods and by studying the electrical properties of the films. The structural perfection of the films remained higher after implantation with more energetic arsenic ions (350 keV vs 190 keV). 100%-activation of implanted ions as a result of post-implantation annealing was achieved, as well as the effective removal of radiation-induced donor defects. In some samples, however, activation of acceptor-like defects not related to mercury vacancies as a result of annealing was observed, possibly related to the effect of the substrate.

1 Introduction

Ion implantation remains to be the most widely used method for fabrication of *p-n* junctions for photodiodes based on $Hg_{1-x}Cd_xTe$ (MCT), the basic photoelectronics material in the infrared part of the spectrum. Nowadays, the most popular photodiode technique is based on the implantation of arsenic, with fabrication of a p^+-type implanted region in an *n*-type material. Dark currents in *p-n* junctions of the 'p^+–n' type are believed to be substantially lower than those in 'n^+–p'-type structures, which provides a higher working temperature or a longer wavelength cut-off sensitivity threshold of photodiodes [1-3]. This is due, among other things, to the fact that *n*-type MCT typically has lower concentration of deep levels which limit the minority carrier lifetime and thus, the performance of the photodetectors.

After arsenic implantation, the implanted MCT region typically has *n*-type conductivity and high density of radiation-induced structural defects. To achieve *p*-type conductivity, the number of the defects should be reduced and arsenic must be activated electrically. For this purpose, activation annealings are performed, but the best implantation and annealing strategy is yet to be found. For example, problems of arsenic activation were reported with the whole epitaxial structure sometimes converting into *p*-type conductivity after implantation and annealing [4].

The purpose of this work was to investigate defect structure of arsenic-implanted MCT films with the use of optical reflectance and Raman scattering as well as via studying the electrical properties of the films.

2 Experimental

In this study, we used $Hg_{1-x}Cd_xTe$ epitaxial structures with composition (CdTe molar fraction) x ranging from 0.23 to 0.30. The structures were grown with molecular-beam epitaxy on Si (013) substrates with CdTe/ZnTe buffer layers [2] and had an "active' layer with constant x and a vary-band protective layer with composition at the surface approaching x=0.45. As-grown structures had *n*–type conductivity with electron concentration as measured at the temperature T=77 K n_{77}=(2.8–7.4)×10^{14} cm^{-3}. They were implanted with As^+ or As^{++} ions with E=190 keV or 350 keV energy, respectively, and 10^{14} cm^{-2} dose. After implantation, the structures were subjected to thermal activation annealing in saturated Hg vapors (360 °C, 2 h and 220 °C, 24 h).

Ion implantation was performed using IMC200 (Ion Beam Services, France) installation and arsenic profiles

* Corresponding author: mynkad@mail.ioffe.ru

were studied using Secondary-Ion Mass-Spectroscopy (Cameca IMS-6F, France).

Optical reflectance spectra were recorded in 300 to 800 nm wavelength range at T=300 K using Shimadzu UV-3600 (Japan) spectrometer.

Micro–Raman spectra were recorded at T=300 K in the spectral range 50–1300 cm^{-1} using Renishaw in Via Raman System with HeNe laser (wavelength 633 nm), a cooled CCD camera and an optical Leica microscope.

Electrical parameters of the films were determined with the measurements of the magnetic field dependence of the Hall coefficient R_H and conductivity σ at T=77 K. The dependences were analyzed using discrete mobility spectrum analysis (DMSA) method [5].

3 Results and Discussion

Figure 1 summarizes Secondary-Ion Mass-Spectroscopy (SIMS) data for the samples studied. As can be seen, experimental arsenic profiles could be grouped into two types, according to the implantation energy used. The experimental profiles could be well fitted with calculated ones obtained with the use of the TRIM software. Average concentration of implanted arsenic was estimated as $N_{as}(avg) = A_{As}/d_r$, where A_{As} is the area under the profile and d_r is implantation depth (0.3 μm for 190 keV ion energy and 0.5 μm for 350 keV ion energy). For E=190 and 350 keV, N_{as}(avg) was 3.7×10^{18} cm^{-3} and 2.2×10^{18} cm^{-3}, respectively.

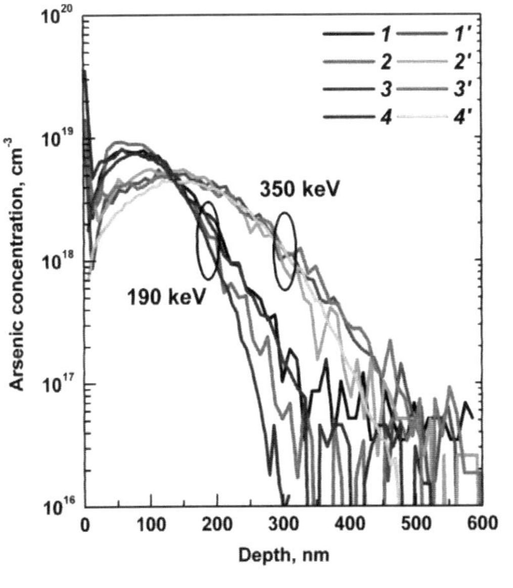

Fig. 1. Experimental arsenic profiles for samples with x=0.30, 0.23 and 0.29 (curves *1*, *1'*; *2*, *2'* and *3*, *3'* respectively), implanted with ion energy 190 keV (curves *1-3*) and 350 keV (curves *1'-3'*), respectively. Curves *4* and *4'* show the results of TRIM modelling.

Figure 2 shows optical reflectance spectra for the sample with x=0.30. It is known that in MCT optical reflectance spectra, there exists a typical doublet of E_1 and $E_1+\Delta_1$ peaks, which are due to transitions $\Lambda_{4,5} \rightarrow \Lambda_6$ and $\Lambda_6 \rightarrow \Lambda_6$, respectively. The position of the peaks provides information on the energy gap (composition x), while the shape of the peaks speaks of the structural perfection of the material [6]. It followed from the reflectance data obtained that the 'sharpness' of the peak E_1 was higher after implantation with As^{++} ions with E=350 keV than after implantation with As$^+$ ions with E=190 keV. This can be clearly seen in Figure 2, similar results were obtained for other samples. This meant that structural perfection at the surface was higher after implantation with more energetic ions.

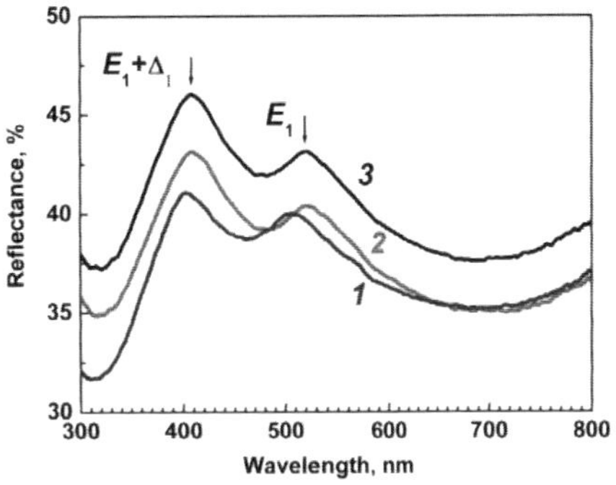

Fig. 2. Optical reflection spectra for the sample with x=0.30: as-grown sample (curve *1*), and samples after implantation with As^{++} 350 keV ions (curve *2*) and As$^+$ 190 keV ions (curve *3*).

For all the samples subjected to ion implantation and activation annealing, the chemical composition at the surface determined according to the position of the E_1 peak appeared to be smaller than that before activation annealing, thus, the annealing led to the enrichment of the surface vary-band protective layer with mercury.

These conclusions were confirmed with micro-Raman spectroscopy data (spectra not shown). According to the identification data available in the literature ([6-11]), typical Raman spectra of the as-grown samples contained lines which could be associated with TA_2, LA_2 (HgTe), LO_1 (CdTe) and LO_2 (HgTe) phonon modes and modes related to Te_{Hg} and V_{Hg} defects and Te nanoinclusions. All the lines remained in the spectra of implanted samples with some re-distribution of intensities.

Figure 3 shows primary envelopes of mobility spectra for the studied sample with x=0.30. Shown are the envelopes for the as-grown sample, sample after ion implantation (energy 350 keV, dose 10^{14} cm^{-2}) and arsenic activation annealing, similar implanted and annealed sample after 1 μm of the surface layer was removed with chemical etching, and the same sample with extra 1.2 μm removed with the etching.

In the as-grown sample, conductivity was mostly determined by active-layer electrons with concentration n_{77}~3×10^{14} cm^{-3} and low-mobility electrons of a transitional layer adjacent to the buffer layer. After implantation and annealing, conductivity increased from 0.36 up to 0.85 (Ohm·cm)$^{-1}$ and was now defined by heavy holes with mobility μ_h=110 cm^2/(V·s), and light holes. High-mobility electrons of the active layer ceased to contribute to conductivity. This could be explained by formation of hole-type conductivity. After chemically removing Δd_I~1 μm, the set of carriers remained the same, however, concentration of heavy holes went down from 4×10^{16} cm^{-3} to 4.2×10^{15} cm^{-3}, while mobility increased from 110 up to 310 cm^2/(V·s), witnessing to removal of surface layer containing implanted (and activated) arsenic. The removal of extra 1.2 μm did not affect the parameters. Thus, in this case activation annealing led to conductivity type conversion across the whole thickness of the structure, which was due to generation of acceptors that most probably were not related to Hg vacancies. Similar observations related to activation, as a result of implantation and annealing, of yet not identified acceptors in MCT epitaxial films grown on Si were reported on the basis of photoluminescence studies performed on the material with x~0.4 [4].

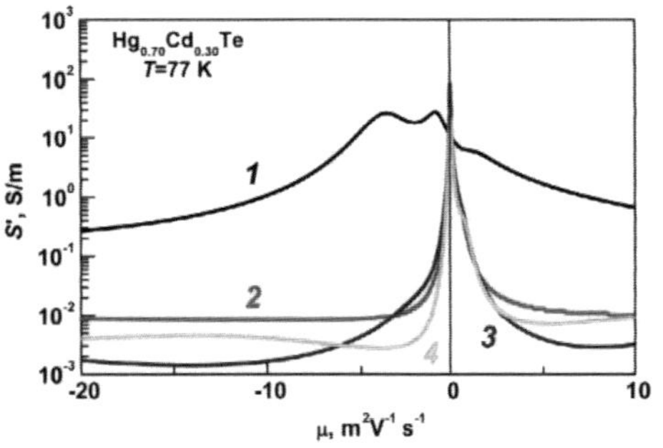

Fig. 3. Primary mobility envelopes for the sample with x=0.30: *1*, as-grown sample; *2*, sample after implantation and annealing; *3*, sample after 1 μm was removed with chemical etching; *4*, sample after extra 1.2 μm was removed with the etching.

Figure 4 shows primary envelopes of mobility spectra for the sample with x=0.23. Presented are the envelopes for the as-grown sample, sample after ion implantation (energy 350 keV, dose 10^{14} cm^{-2}) and arsenic activation annealing, similar implanted and annealed sample after 0.5 μm of the surface layer was removed with chemical etching, and the same sample after extra 0.9 μm was removed with the etching.

In as-grown sample conductivity σ was mostly determined by active-layer electrons with concentration n_{77}~$8.6\cdot10^{14}$ cm^{-3} and low-mobility electrons of the transitional layer at the interface with the CdTe buffer layer, typical of such type of structures [2]. After implantation and annealing, σ decreased from 3.7 down to 2.8 (Ohm·cm)$^{-1}$ and was now defined by heavy holes with mobility μ_h=32 cm^2/(V·s), light holes and active-layer and transitional-layer electrons. This can be explained by formation of hole-type conductivity. After chemically removing Δd_I~0.5 μm, the set of carriers changed as heavy holes have gone: thus, the surface layer containing implanted (and activated) arsenic was removed. The removal of extra 0.9 μm-thick layer did not affect the parameters. Thus, activation annealing in this case led to activation of implanted arsenic across the thickness of the implanted layer and also modified the properties of the rest of the active layer (which remained n-type). A similar result was obtained for the MCT sample with x=0.29.

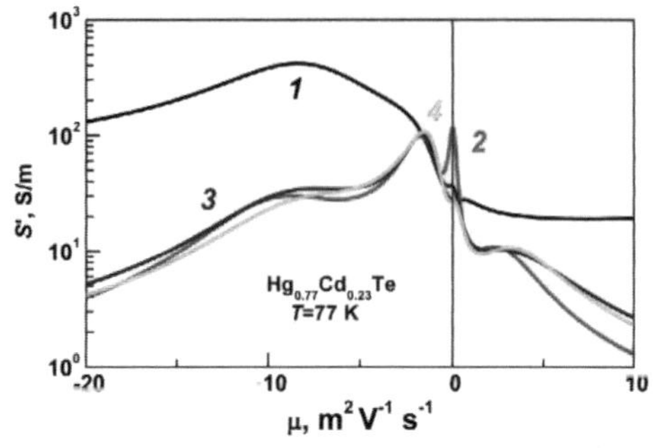

Fig. 4. Primary mobility envelopes for the sample with x=0.23: *1*, as-grown sample; *2*, sample after implantation and annealing; *3*, sample after 0.5 μm was removed with chemical etching; *4*, sample after extra 0.9 μm was removed with the etching.

On the basis of SIMS data and results of electrical measurements, we were able to make a comparison of 'real' heavy hole concentration in implanted layer $p_r=p_h\cdot\Sigma d/d_r$ with average arsenic ion concentration $N_{As}(avg)$ mentioned above. For example, for the sample with x=0.23 implanted with 350 keV ion energy, $p_r=p_h\cdot\Sigma d/d_r=2.25\times10^{17}\times 8.3/0.5=3.74\times10^{18}$ cm^{-3}, where Σd is the total thickness of MCT film. N_{As}(avg) for this sample was $2.2\cdot10^{18}$ cm^{-3}.

The results of the comparison are summarized in Table 1. As can be seen in this Table, heavy hole concentration in implanted layer agrees quite well with average arsenic ion concentration, with sample with x=0.29 implanted with 350 keV ion energy being the only exception. Thus, for these samples we can speak of 100% arsenic activation as a result of annealing.

Table 1. Electrical activation of the implanted arsenic.

x	0.23	0.23	0.29	0.29
Σd, μm	8.3	8.3	7.4	7.4
Ions	As^+	As^{++}	As^+	As^{++}
p_{77}, cm^{-3}	$9.3\cdot10^{16}$	$2.3\cdot10^{17}$	$1.8\cdot10^{17}$	$4.7\cdot10^{16}$
μ_{p77}, cm^2/(V·s)	66	32	19	77
p_{r77}, cm^{-3}	$2.6\cdot10^{18}$	$3.7\cdot10^{18}$	$4.4\cdot10^{18}$	$6.9\cdot10^{17}$
d_r, nm	300	500	300	500
N_{As}(avg), cm^{-3}	$3.7\cdot10^{18}$	$2.2\cdot10^{18}$	$3.7\cdot10^{18}$	$2.2\cdot10^{18}$

4 Conclusions

In this work, the defect structure of arsenic-implanted HgCdTe films with the use of optical reflectance and Raman scattering as well as via studying the electrical properties of the films was investigated. The studies of optical reflection showed that the structural perfection of the samples remained higher after implantation of more energetic arsenic ions (350 keV vs 190 keV). In the samples with the composition of the active layer x=0.23 and x=0.29, as a result of implantation and activation annealing, a 'p^+–n'-type structure was formed. The concentration of heavy holes in implanted area reduced to the thickness of arsenic profile agreed quite well with the concentration of implanted ions. This spoke of 100% of activation of implanted ions as a result of annealing. Indeed, in this case activation annealing also removed radiation-induced defects of donor type.

In the sample with x=0.30, the annealing led to the formation of a 'p^+–p'-type structure. The conductivity type conversion throughout the whole thickness of the structure in this case was most probably caused by the generation (activation) of acceptor-like defects not related to mercury vacancies.

Generally, optical reflectance, micro-Raman studies and electrical measurements were shown to provide highly detailed characteristics of the implanted structures based on MCT.

The researchers at A.V. Rzhanov Institute of Semiconductor Physics acknowledge the support from the Ministry of Education of the Russian Federation (Grant RFMEFI60414X0134).

References

1. L. Mollard, G. Bourgeois, C. Lobre, S. Gout, S. Viollet-Bosson, N. Baier, G. Destefanis, O. Gravrand, J.P. Barnes, F. Milesi, J. Electron. Mater. **43**, 802 (2014)
2. M.V. Yakushev, V.S. Varavin, V.G. Remesnik, D.V. Marin, Semiconductors **48**, 767 (2014)
3. A.M. Itsuno, P.Y. Emelie, J.D. Phillips, S. Velicu, C.H. Grein, P.S. Wijewarnasuriya, J. Electron. Mater. **39**, 945 (2010)
4. K.D. Mynbaev, N.L. Bazhenov, M.V. Yakushev, D.V. Marin, V.S. Varavin, Yu.G. Sidorov, S.A. Dvoretsky, Tech. Phys. Lett. **40**, 708 (2014)
5. V.V. Bogoboyashchyy, A.I. Elizarov, I.I. Izhnin, Semicond. Sci. Technol. **20**, 726 (2005)
6. S.A. Dvoretsky, N.N. Mikhailov, V.G. Remesnik, N.Kh. Talipov, Avtometriya **5**, 73 (1998) (in Russian)
7. A.I. Belogorokhov, I.A. Denisov, N.A. Smirnova, L.I. Belogorokhova, Semiconductors **38**, 84 (2004)
8. A.I. Belogorokhov, N.A. Smirnova, I.A. Denisov, L.I. Belogorokhova, B.N. Levonovich, Phys. Stat. Sol. C **7**, 1624 (2010)
9. I. Hill, J. Appl. Phys. **67**, 4270 (1990)
10. S.H. Shih, J. Bajai, L.A. Moudy, D.T. Cheung, Appl. Phys. Lett. **43**, 68 (1983)
11. A. Tverjanovich, K. Rodionov, E. Bychkov, J. Sol. State Chem. **190**, 271 (2012)

Dirac's HdCdTe semimetals grown by MBE technology

Jakub Grendysa[*,1], *Charles R.*Becker[1], *Malgorzata* Trzyna[1], *Renata* Wojnarowska-Nowak[1], *Ewa* Bobko[1,2] and *Eugen M.* Sheregii[1]

[1]Centre for Microelectronics and Nanotechnology, University of Rzeszow, Pigonia 1, 35-959 Rzeszow, Poland
[2]Institute of Physics, Polish Academy of Sciences, Lotników 32/46 Av., 02-668 Warszawa, Poland

Abstract. Peculiarities of the MBE growth technology for the Dirac's semimetal based on the $Hg_{1-x}Cd_xTe$ alloys have been presented. Composition of layers was controlled by ToF-SIMS, FTIR measurements, and by the $E1+\Delta1$ maximum position of optical reflectivity in visible reason. The surface morphology has by determined via atomic force and electron microscopy.

1 Introduction

Since 1958 the HgCdTe alloys are main material for infrared devices [1]. In last decade (from 2007 [2]]) a new property of this material, namely, inverted band structure caused by strong spin-orbital interaction, forms a new opportunity for this material – possibility to produce topological phase on surface. By this way, the Topological Insulators (TI) based on the HgCdTe became actual in 3D case [3].

In this short communication is presented peculiarity of the MBE technology designed to the strong TI producing based on semi-metallic $Hg_{1-x}Cd_xTe$ Te with $x\leq 0.16$ what is semimetal yet but close to the 3D Dirac point .

2 Experiment

2.1 MBE Grown

The Dirac's semimetal $Hg_{1-x}Cd_xTe$ epilayer samples has been grown in MBE system Riber Compact 21 with CdTe, Te, ZnTe, Hg, In standard cells and CdI_2 cells with a special insert on the top to limit the value of the I flux. The $Hg_{1-x}Cd_xTe$:I we received using Te, CdTe, Hg, and CdI2 source. Two type of substrates were used) GaAs substrates (001) oriented were applied with the MBE grown CdTe buffer that produced ~ 0.3% mismatch with the next thin MBE grown $Hg_{1-x}Cd_xTe$ layer and CdZnTe substrates (112)B orientated were employed and the $Hg_{1-x}Cd_xTe$ was grown directly on the CdZnTe(112)B substrate surface which have a lattice constant practically (with an uncertainty of 0.01%).

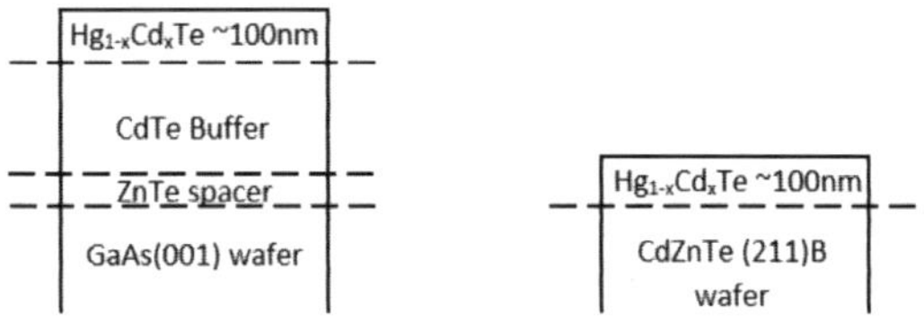

Fig. 1 The structure of the samples

Several conditions should be fulfilled during the growth process of the $Hg_{1-x}Cd_xTe$ layer with $0<x<0.16$ of high quality – with surface close to ideal and low concentration of defects (hillocks).

At the very beginning we had to carefully prepare a substrate. Because they were used two types of substrate (See Fig. 1) were different ways to prepare them: i) for the GaAs substrate (001) was enough degassing at 350°C and then deoxydation of the temperature of about 580°C. ii) for the first CdZnTe substrate was rinsed in acetone, methanol, then a solution of 1% bromine in methanol [4] and dry nitrogen gas. Then, degassing at a temperature of 130°C (the temperature is lower as compared to those of GaAs substrate CdZnTe assembled on the carrier by indium), and oxygenate at 300°C (over 250°C above the point the flux on the substrate Te).

The substrate temperature should be controlled with accuracy not worthy then ±0.5°C because its strong influence on composition [5–7]. The substrate temperature determination need of specific calibration due to lack of physical contact for thermocouple with molybdenum substrate holder. Therefore, this thermocouple is calibrated by using the melting point of In, Sn and PbSn and PbAg on holder.

Next step is the determination and stabilization of the beam equivalent pressure (BEP) of sources is principally important at MBE growth for each compounds. In case

* Corresponding author: grendysa@ur.edu.pl

of the $Hg_{1-x}Cd_xTe$ layer growth additional factor is essential: stabilization of the Hg-flux is difficult and long – for two hours long [8]. The composition is determined by the Cd/Te flux ratio whereas, the flux ratio Hg/Te is important for quality of received layers [5,8]. Stabilization of these two ratio at the growth is crucible: deviation of BEP in the framework of 10^{-8} Torr able to change considerably the composition in the case of the Cd/Te ratio as well as to increase the hillock density, electron concentration and decrease the electron mobility in the case of the Hg/Te ratio [5]. To stabilize latter one we need two ours long operation of the sources temperatures. Simultaneously, the choose and stabilization of the flux parameters is complicated and it is necessary to prepare special computer program enables us to find corresponding flux values for Te and Hg when the CdTe fluxe is introduced: if the CdTe flux for the case of x= 0.15 has value $2{,}9\cdot10^{-7}$ then the Te and Hg ones are correspondently chosen.

Last and very important step is the temperature regime of substrate after the end of process is also important in the case of the $Hg_{1-x}Cd_xTe$ layer growth: it is necessary to leave the sample in atmosphere of mercury until the temperature of mountain decreases lower than 100°C in order to prevent the evaporation of Hg from surface.

As was mentioned above, accurate calibration of the BEP is cruel (particularly for x equal to about 0,15) and was performed by two steps: before insert the GaAs substrate into the chamber and at starting of the growth process. To stabilise the fluxes of Te, CdTe and Hg it is necessary to avoid of sudden changes of its values: such calibration should be made before any increase in order. Once known flux is what we want to get, (after initial calibration carried out immediately after loading of source materials and degassing main chamber MBE), it is possible to choose it prior to narrow down the temperature range for the interesting range.

Table 1 Parameters of strained and no strained Epitaxial Layers $Hg_{1-x}Cd_xTe$

	Composition, x, mol.	Substrate	Buffer	Thickness, nm
MCT 32	0.157	GaAs (001)	ZnTe/ CdTe	590
MCT 37	0.135	CdZnTe (112)*B*	-	100
MCT 41	0.135	GaAs (001)	ZnTe/ CdTe	100
MCT 42	0.155	GaAs (001)	ZnTe/ CdTe	100
MCT 44	0.135	GaAs (001)	ZnTe/ CdTe	100

It is important to start the growth of the fresh CdTe layer, preferably directly prior to actual growth. When the oxide layer is removed (the RHEED pattern is appeared) the PSCT temperature should be immediately decreases to 300°C and start with the CdTe buffer growth. The first was the layers of ZnTe (about 0.1 μm) as a small crystal lattice mismatch with respect to GaAs than the CdTe (3.5μm layer is applied [9]).

In case of CdZnTe substrates the layer of HgCdTe we are grown directly at cleaned and deoxidation surface.

The next step is the opening of the valve spool of Hg and introduction of mercury into the main chamber by raising out of the external tank using elevators and heating Hg cell to a temperature at which we get the desired flux. It is a very long and important process, because the accuracy of the temperature settings strongly influence the fluxes and thus the Hg/Te ratio what is important for "quality" of received layers [8,10].

2.2 Controlled of layers composition

Composition of the resulting layers was controlled by Time-of-flight secondary ion mass spectrometry (ToF-SIMS)[11] data were acquired using a ION-TOF GmbH TOF-SIMS V spectrometer. The analysis chamber was maintained at less than 5×10^{-10} Pa under operational conditions. Sputtering was performed using the 2 keV (80 nA) Cs+ ion beam, rastered over an area of 300×300 μm. A pulsed 30 keV Bi+ primary ion source at a current of 1.3 pA. (high current bunched mode), rastered over a scan area of 100×100 μm was used as the analysis beam. TOF-SIMS depth profiles were measured with the instrument working in the dual-beam mode.

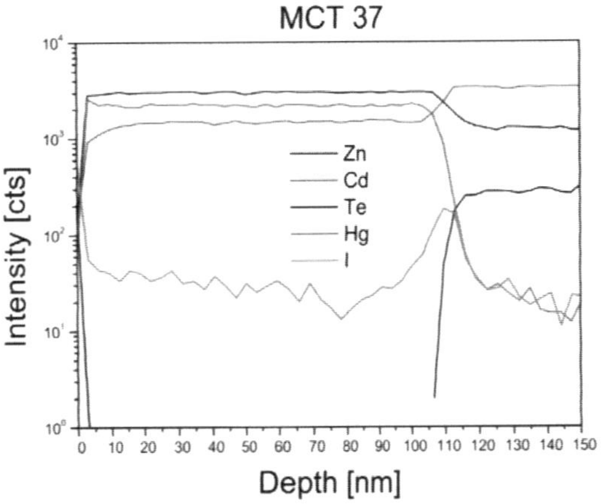

Fig. 2 Profile the downhole – Measurment ToF-SIMS of MCT37

The AFM measurements were performed using a AFM Inova Brucker in Taping Mode area 0,5-10μm at 512 line with 0,1-0,5Hz

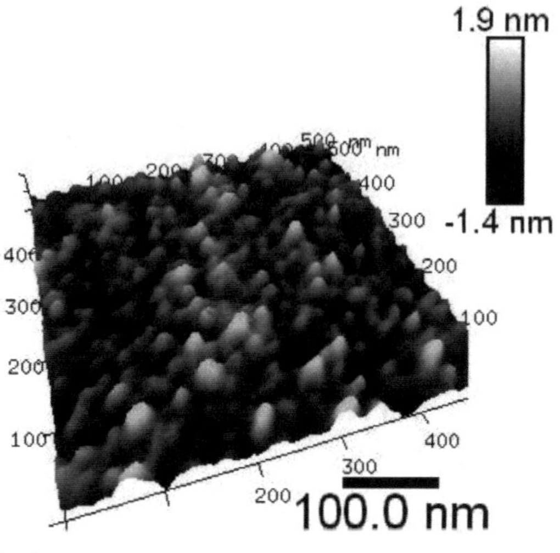

Fig. 3 AFM Hight resolution image of sample MCT37

The optical properties were studied by Fourier transform infrared (FTIR) spectroscopy method. The measurements were performed by Vertex 70v (Bruker) spectrometer, in the transmission mode. The FTIR spectra were obtained in the region of middle infrared (400 – 7500 cm^{-1}), using 32 scans with spectral resolution of 2 cm^{-1}.

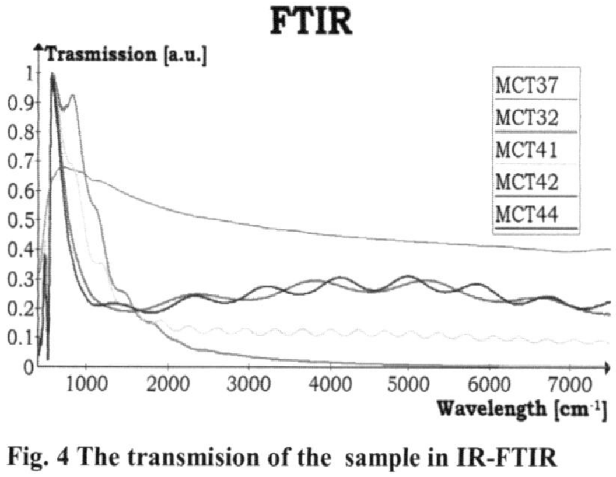

Fig. 4 The transmision of the sample in IR-FTIR measurements.

Discussion and Conclusion

In summary, more than forty layers have been grown by described above method. A quantized Hall conductance in the 3D $Hg_{1-x}Cd_xTe$ samples ($x < 0.16$) with thickness from 100 to 1280 nm and with different levels of n type doping, using either iodine or indium as dopants, is observed over a wide temperature region from 0.4 to 50 K. In case of sample grown on GaAs/CdTe substrates – MCT 37, 42, 44 – the clear Quantum Hall Effect was observed with quantized values of the Hall resistances in plateau equal to $R_{xy} = h/\nu e^2$ (ν is one integer) which was accompanied zero R_{xx}. That means the strong TI was realized [6] enabled by applied effective MBE technology.

Acknowledgments

We acknowledge support from the authorities of the Podkarpackie Voivodship (Marshals Office of the Podkarpackie Voivodship of Poland), contract WND-RPPK.01.03.00-18-053/12.

References

1. R. Dornhaus and G. Nimtz, *Narrow Gap Semiconductors* **98**, 159 (Springer-Verlag, Berlin, 1985)
2. M. König, S.Wiedmann, C. Brune, A. Roth, H. Buhmann, L. W. Molenkamp, Xiao-Lian Qi and S.-C. Zhang, *Science* **318**, 766–770 (2007)
3. G. Tomaka , J. Grendysa, P. Śliż, C. Becker, J. Polit, R. Wojnarowska, A. Stadler and E.M. Sheregii, *Phys. Rev. B*. **93**, 205419-7 (2016)
4. J. Zhao, Y. Chang, G. Badano, S. Sivananthan, J. Markunas, S. Lewis, J. H. Dinan, P. S. Wijewarnasuriya, Y. Chen, G. Brill, and N. Dhar, J. Electron. Mater. **33**, 881 (2004)
5. L. He, C. R. Becker, R. N. Bicknell-Tassius, S. Scholl, and G. Landwehr, J. Appl. Phys. **73**, 3305 (1993)
6. J. Bajaj, J. M. Arias, M. Zandian, D. D. Edwall, J. G. Pasko, L. O. Bubulac, and L. J. Kozlowski, J. Electron. Mater. **25**, 1394 (1996)
7. F. Goschenhofer, J. Gerschütz, A. Pfeuffer-Jeschke, R. Hellmig, C. R. Becker, and G. Landwehr, J. Electron. Mater. **27**, 532 (1998)
8. J. Schmidt, K. Ortner, J. E. Jensen, and C. R. Becker, J. Appl. Phys. **91**, 451 (2002)
9. L. He, S. Wang, J. Yang, M. Yu, Y. Wu, X. . Chen, W. Fang, Y. Qiao, Y. Gui, and J. Chu, J. Cryst. Growth **201**, 524 (1999)
10. Y. Chang, C. R. Becker, C. H. Grein, J. Zhao, C. Fulk, T. Casselman, R. Kiran, X. J. Wang, E. Robinson, S. Y. An, and Others, J. Electron. Mater. **37**, 1171 (2008)
11. L. Wang, L. H. Zhang, and J. Li, J. Electron. Mater. **28**, 793 (1999)

Growth and characterization of organic layers deposited on porous-patterned Si surface

Tamara Ya. Gorbach[1], *Petro S.* Smertenko[1], *G.P.* Olkhovik[1], *Grzegorz* Wisz[2]

[1]V.Lashkarev Institute of Semiconductors Physics, NAS of Ukraine, 45, Prospekt Nauki, 03028, Kyiv, Ukraine
[2]Chair of Experimental Physics, Faculty of Mathematics and Natural Sciences , University of Rzeszow, S. Pigonia 1, 35-959 Rzeszow, Poland

Abstract. The organic layers with the thickness from a few nanometers up to few micrometers have been deposited from the chemical solution at room temperature on porous patterned Si surfaces using two medical solutions: thiamine diphosphide (pH=1÷2) and metamizole sodium (pH=6÷7). Based on evolution of morphology, structural and compositional features obtained by scanning electron microscopy, X-ray analysis, reflectance high energy electron diffraction the grown mechanisms in thin organic layers are discussed in the terms of terrace-step-kink model whereas self-organized assemblies evaluated more thick layers. Transport mechanism features and possible photovoltaic properties are discussed on the base of differential current-voltage characteristics.

1 Introduction

Organic materials – pigments, dyes, polymers and their combination with the typical inorganic ones such as metals, metal oxides, semitransparent or semiconductor substrates have been intensively investigated over the last decades []. Currently hybrid organic-inorganic nanocomposites are considered as materials being suitable for a wide range of applications [1]. The electroluminescent devices [2-4], thin field effect transistors [5, 6], solar cells [7- 11], etc. were prepared, characterized and reported. Due to the nanosized molecules of organic materials as well as the low-dimensional semiconductor systems the novel and the special properties of thin organic layers and hybrid formations are observed. They are proposed as the next generation for electronics devices and potential candidates for practical application in different fields of electronics, optoelectronics, photovoltaics, biosensors, biomedicine, etc.

Various technological methods such as thermal or laser evaporation, chemical vapor deposition, chemical solution growth, spin coating and many others were developed for preparing the active organic multilayer. Information about growth mechanisms, surface and interface morphology, crystalline structure, chemistry, composition of thin layer is prerequisite to understand and fabricate their precise and desirable properties. In our previous work [12] it was shown that improvement of initial *Si*-based solar cell with EFF less that 7% up to 15% can take place by treatment in organic solutions. As organic materials the medical preparations were chosen. The *Si* n^+p

DOI: 10.1051/epjconf/201713301003

patterned solar cells were the waster of technology taking place in photovoltaics.

This paper presents some experimental data on the porous patterned *Si* structures stimulated still much investigation for application in optoelectronics and photovoltaics.

2 Experimental

The deposition of organic layers onto porous patterned *Si* substrate was carried out from aqueous solution of thiamine diphosphide (TD) and metamizole sodium (MS) at room temperature during 100 min and 20 hours. These organic materials are medical ones. They were deposited onto Si surface without preliminary purification. The molecular structure of organic medical material is shown in Fig.1.

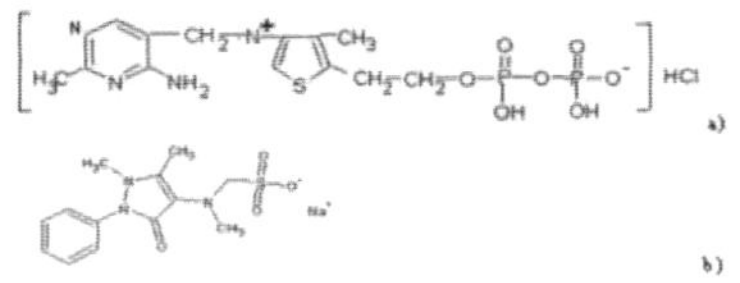

Fig. 1. Molecular structure of the organic materials: a) – thiamine diphosphide, contained *H, C, N, O, S, P, Cl* elements; b) – metamizole sodium, contained *H, C, N, O, S, Na* elements.

According to chemical formula the first material contains *H, C, N, O, S, P, Cl* elements, has an acid chemistry (pH=1±2). It is known in medicine as cocarboxylase hydrochloride. The second one has alcaline chemistry (pH=6±7), contains *H, C, N, O, S, Na* elements and named analgin.

The porous Si was formed by electrochemical etching (ECE) in $HF:C_3H_7OH$=1:1 electrolyte at current density 10-30-mA/cm^2 on patterned pyramid *Si* like surface. The last was produced on {100} cutting, 10 Ωcm resistivity, B-doped CZ-*Si* wafers with thickness 300 μm by anisotropic chemical etching (ACE) in 10% *KOH* at room temperature during some minutes. Patterned porous surface has average pyramid dimension of 2-3 μm at the base and 1.5-2 μm in the height with the thickness of porous layer is about 0.2-0.3 μm.

Superprobe microanalyzer JXA-733 and optical microscopy were used to obtain information about morphology features, composition and growth mechanisms of deposited layers. The layer crystallinity was characterised by reflection high-energy electron difraction (RHEED) with electron microscopy JEM-6A.

The current-voltage characteristics (I-VC) were measured in *In*-organic-PS-*Si*-*In* sandwich systems in the dark and under illumination in both directions using automated tester 14 TKS-100 (USSR). For illumination the Xenon lamp was used with intensity equivalent 0.03 sun. Analysis of *I-V* characteristics of hybrids in the terms of differential slope $\alpha = dlogI/dlogV$ was carried out [13, 14].

3 Results and discussion

3.1 Morphology

The typical morphology fragments of layers deposited from TD and MS solutions are presented in (Fig.2 a, c-e) and (Fig.2 f-h), respectively. Fig. 2b shows RHEED pattern observed after layer deposition from TD solution during 100 min. As seen from Fig.2 there are no significant effect on organic layer morphology formed from solution with different chemistry. On early stage formation the layer contour is the same as the substrate one. There is the morphological stability condition during growth process. Due to excellent morphological feature of facet {111} which is a vicinal surface and contains a quasiperiodic flat low-Miller- index terraces separated by steps with kinks the surface midiated heteroepitaxy take place.

 DOI: 10.1051/epjconf/201713301003

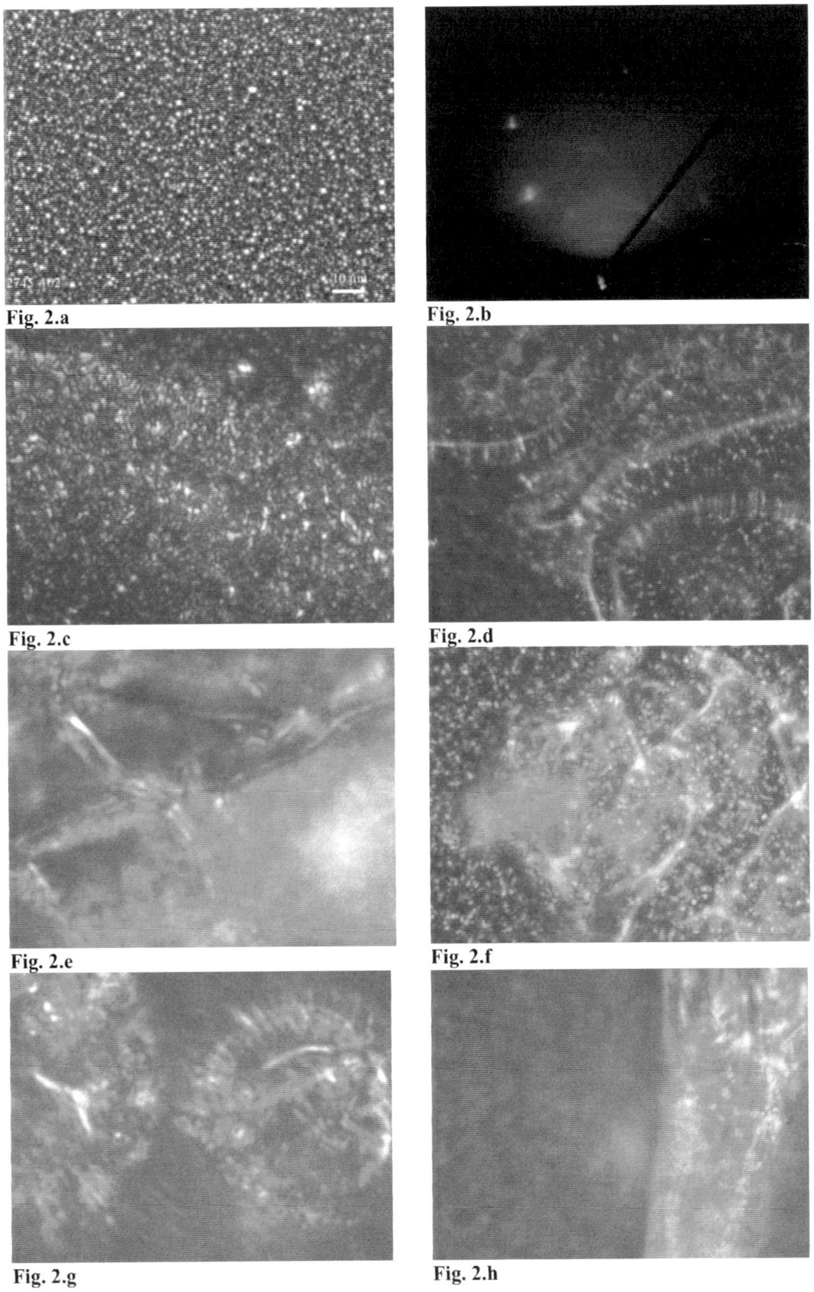

Fig. 2. a, b – morphology and RHEED pattern of layer deposited from TD solution during 100 min; **c-e** and **f-h** – morphology of layers deposited from TD and MS solution during 20 H, respectively.

 DOI: 10.1051/epjconf/201713301003

In such case a low temperature and even room temperature deposition of good quality layer is possible in spite of the factor of the big lattice mismatch and difference in the thermal expansion coefficient [15]. The scale of the roughness prepared by ECE is essentially less than TSK feature size. That is why the TSK growth model is realized. The RHEED observation (Fig.2b) shows the spots and Kikuchi lines. In this RHEED pattern the high quality heteroepitaxy is testified.

With deposition time increasing till 20 hours the loss stability conditions take place and complex morphologies are grown in circular, square and rhombic forms. These forms include mostly filaments and dots (Fig. 2 d-h). However these formations are not only due to self-organisation. In this case the substrate pre-pattern determines their further development too.

According to X-ray analysis the both deposited layers exhibit the same components as solution. Because TD contains often *K, Na, Ca,* Mg ligands the X-ray analysis of thin layers was shown the presence of *Ca* content and *Na* element in thick layers. These elements are absent in molecular formula (Fig.1a).

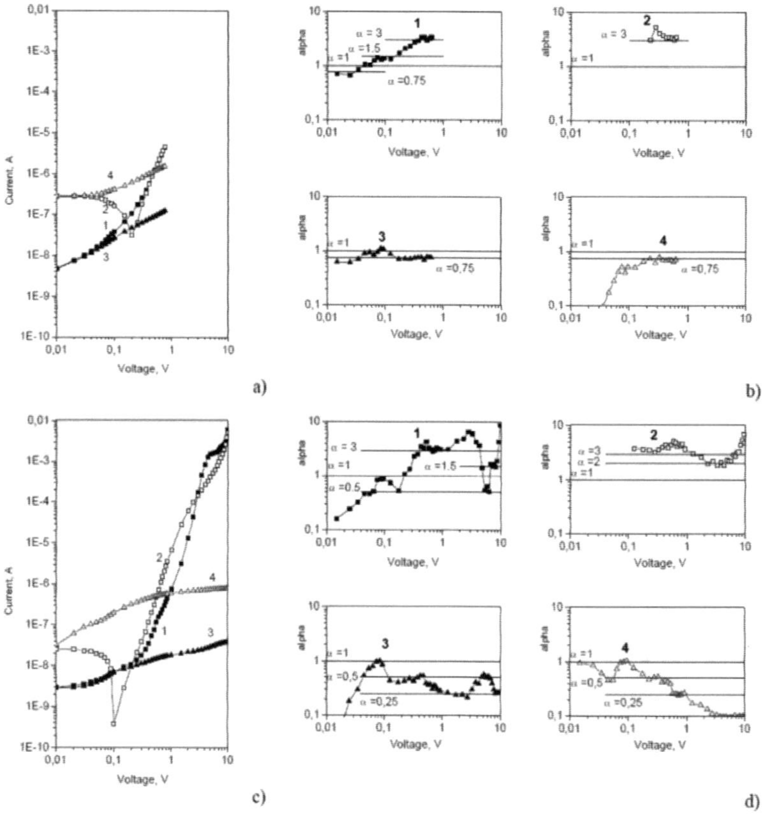

Fig. 3. Current-voltage characteristics a), c) and their differential image b), d) for the layers deposited from TD during 100 min – a), b) and during 20 hours – c), d); curve 1 forward dark, curve 2 forward under illumination, curve 3 reverse dark and curve 4 reverse under illumination.

3.2 Current-voltage characteristics

The current-voltage characteristics of the structures under investigation and their differential image are shown in Fig.3 and Fig.4 for TD and MS solutions, respectively. The main feature of all I-VC is the wide variety of differential slope: from sub-linear behaviour with α=0.1 (Fig.3d curve 4); α=0.25 (Fig.3d curves 3, 4); α=0.5 (Fig.3d curves 1, 3, 4); α=0.75 (Fig.3b curves 1, 3, 4; Fig.4d curves 3, 4); to super-linear one with α=1.5 (Fig.3b curve 1; Fig.3d curve 1; Fig.4b curves 3, 4; Fig.4d curves 1, 2, 3); α=2.0 (Fig.3d curve 2; Fig.4b curves 1, 2, 3; Fig.4d curves 1, 2, 3, 4); α=3.0 (Fig.3b curves 1, 2; Fig.3d curves 1, 2; Fig.4b curves 1, 2; Fig.4d curves 1, 2).

Some of such behaviour is well known from literature [16-19]: α=3.0 corresponds to the high injection of charge carriers into dielectric; α=2.0 can be referred to the space charge limited currents or to the double injection under monomolecular recombination of charge carriers with predominance of only one charge carriers type, whereas α=1.5 is occur under bimolecular recombination of charge carriers with approximately equal quantity of both charge carriers types. The unusual behaviour with α=0.25 and α=0.75 we can see for reverse direction for both structures. It can be explained on the base of general speculation that some additive process with law $I \sim V^{0.25}$ influences the I-VC. For Ohm law such influence gives α=0.75 or 1.25. It modifies the rectification behaviour with α=0.5 to α=0.25 or α=0.75. The I-VC with α=1.25 was reported for some organic and amorphous materials [14,20].

It is necessary to note also the other features of the structures under investigation:

1. Asymmetric characteristic for TD solution. The ratio of forward to reverse currents is about 40 at V=0.75 V for the layer under deposition during 100 min in TD solution. It decreases to 6 under deposition during 20 hours. To check the exponential behaviour the following procedure on the base of differential approach [14] was used. For usual exponential law for diode we have $I \sim \exp\{eV/nkT\}$, where e is the electron charge, k is the Boltzman constant, T is absolute temperature and n is the nonideality coefficient. In this case the differential slope $\alpha = eV/nkT$ and at V=0 α=0 (Fig, 5b, curve 1, $\alpha_2 = 19*V$) and estimation for nonideality coefficient gives n=2. The second range of experimental curve can be approximated by equation $\alpha_1 = 6.3(V+0.1)$. For I-VC it gives after integration: $I = AV^b * \exp(BV)$ with $A = 8*10^{-8}$; b=0.63; B=6.3. Comparison of this approximation with experimental data is presented in Fig.5a and b.

The I-CV is described by exponent enough well in all voltage range, but the differential slope shows that after 0.4 V the behaviour is not exponential and α=3.

2. The photosensitivity of the layers deposited from TD and MS solutions. The sensitivity of structures with TD is higher than with MS and in both cases it decreases with increase of layer thickness.

3. Photovoltaic properties for TD solution. The increase of the layer thickness leads to decrease of open circuit voltage and short circuit current.

 DOI: 10.1051/epjconf/201713301003

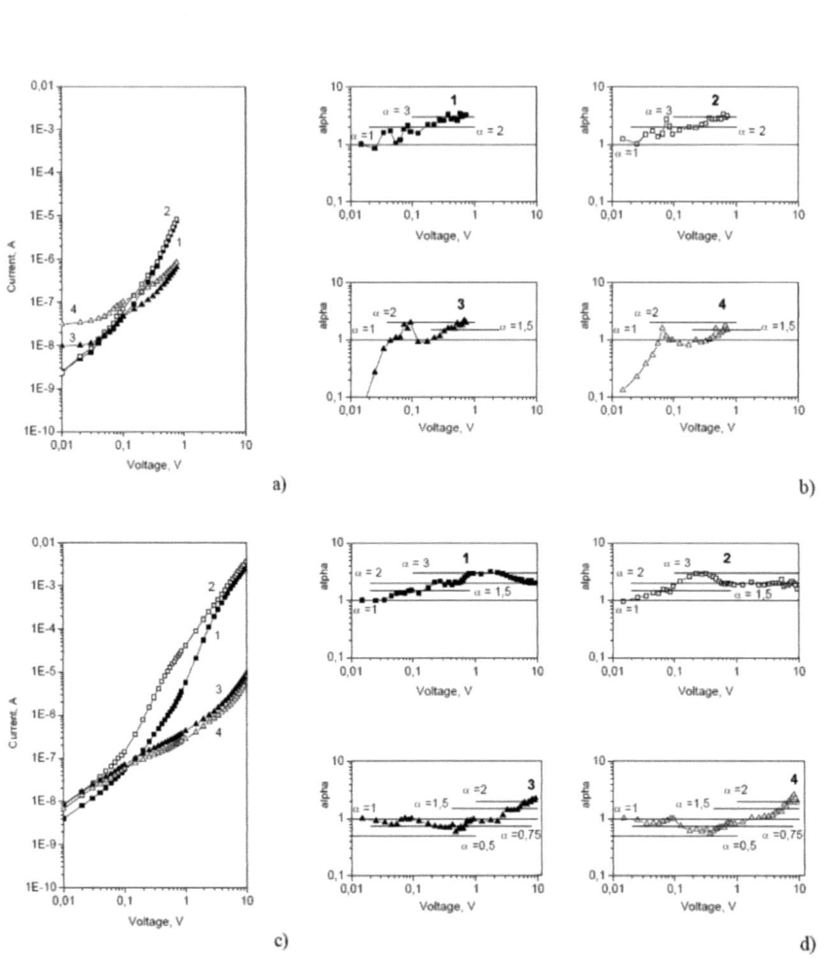

Fig. 4. Current-voltage characteristics a), c) and their differential image b), d) for the layers deposited from MS during 100 min – a), b) and during 20 hours – c), d); curve 1 forward dark, curve 2 forward under illumination, curve 3 reverse dark and curve 4 reverse under illumination.

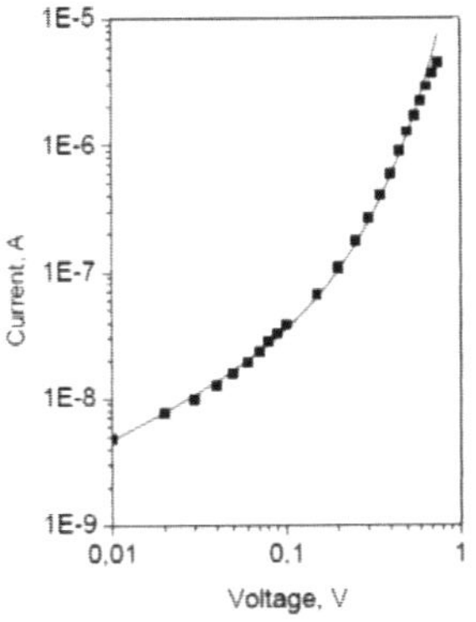

a)

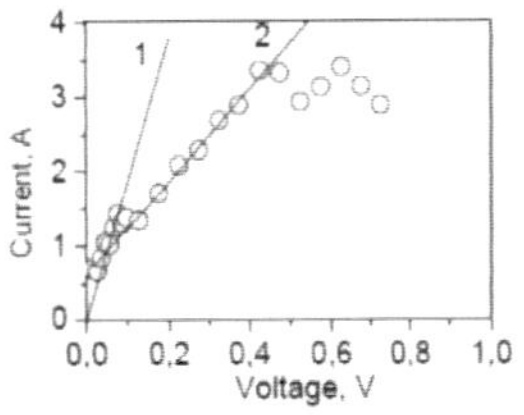

b)

Fig. 5. Modelling of current-voltage characteristic a) and their differential image b) for the layer deposited from TD during 100 min, curve 1 forward dark. Solid curve in a) corresponds to equation: I=AV^b·exp(BV) with A=8·10^{-8}; b=0.63; B=6.3. Solid curves 1 and 2 in b) correspond to equations α_1=19·V and α_2=6.3(V+0.1), respectively.

4 Conclusions

1. Organic layers deposited at room temperature from aqueous solution of medical materials of thiamine diphosphide and metamizole sodium modified the morphology, chemical composition and physical properties of porous patterned substrates. This modification do not depend on solution chemistry and is determined by deposition time, i.e. layer thickness. For thin organic layers (≤100μm) the layer contour is the same as substrate and terrace-step-kink growth mechanism is realized. For more thick layers (up to micrometer) self-organized assemblies are formed but with preference of pre-pattern substrate.
2. On the interface of TD layer – porous patterned substrate the barrier is formed and PV characteristic is observed.
3. In the TD (MS) layer – porous patterned system the transport is determined some processes but mainly the high injection of charge carriers in dielectric organic layers.
4. Chemical solution deposition at room temperature is newly simple technological process useful for realization of hybrid organic-inorganic structures.

References

1. Milliron D. J., Gur I., Alivisatos A. P. MRS Bull. 30, 41-44 (2005)
2. Kyu-Seong Hwang, Ju-Hyun Jeong, Kyung-OK Jeon, Young-Sun Jeon, Byung-Hoon Kim. Ceramics International **33**, 6, 1071-1074 (2007)
3. Gunes S., Sariciftci N.S. Inorg. Chem. Acta **361**, 581-585 (2008)
4. T. W. Baehr-Jones and M. J. Hochberg. J. Phys. Chem. C **112** (21), 8085–8090 (2008)
5. Tao F. Bernasek, S.L. Xu Guo-Quin, Chem. Rev. **109**, 3991-4024 (2009)
6. H. Ma, H.-L. Yip, F. Huang et al., Adv. Funct. Mater. **20**, 9, 1371–1388 (2010)
7. S. Wie, Q. Wang, J. Zhu et al. Nanoscale **3**, 4474–4502 (2011)

8. R. Ding, T. Baehr-Jones, W.-J. Kim et al. J. Lightwave Technol. **29**, 8, 1112–1117 (2011)
9. Alloatti L., Bogaerts W., Dalton L. et al. Journal of Lightwave Technology, **34**, 2, 256-268 (2015)
10. Yuqiang Liu, Zhi-Guo Zhang, Zhouhui Xia et al. ACS Nano 10, 1, 704-712 (2016)
11. M. Lira-Cantú, P, Gómez-Romero. Capter 7 in "*Hybrid nanocomposites for nanotechnology: electronic, optical, biomedical applications*", Ed. Lhadi Merhari (Springer Science+Business Media LLC, 306-308, 2009)
12. J.H. Burroughes, D.D.C. Bradley, A.R. Brown, R.N. Marks, K. Mackay, R.H. Friend, P.L. Burns and A.B. Holmes, Nature **347**, 539 (1990)
13. B.W. D'Andrade, M.A. Baldo, C. Adachi, J. Brooks, M.E. Thompson and S.R. Forrest, Applied Physics Letters **79**, 1045 (2001)
14. T. van Woudenbergh, J. Wildeman, P.W.M. Blom, J.J.A.M. Bastiaansen and B.M.W. Langeveld-Voss, Advanced Functional Materials **14**, 677 (2004).
15. R.W.J. de Boer, M.E. Gershenson, A.F. Morpurgo, V. Podzorov, Phys. stat. sol. (a) **201**, 1302 (2004)
16. S. Scheinert and G. Paasch, Phys. stat. sol. (a) **201**, 1263 (2004)
17. A. Goetzberger, C. Hebling, H.W. Schook, Mater. Sci. Eng. **R40**, 1 (2003)
18. S.E. Shaheen, D.S. Ginley and G.E. Jabbour, MRS Bulletin **30**, 10 (2005)
19. D.J. Milliron, I. Gur, A.P. Alivisator, MRS Bulletin **30**, 41 (2005)
20. H. Hoppe, N.S. Sariciftci, J. Mater. Chem. **16**, 45 (2006)
21. J.C. Hindson, Z. Saghi, J.-C. Garrido Hernandez, P.A. Midgley, N.C. Greenham, Nano Lett. **11**, 2, 904–909 (2011)
22. T.Ya. Gorbach, P.S. Smertenko, S.V. Svechnikov, M. Kuzma, Thin Solid Films **511-512**, 494 (2006)
23. R. Ciach, Yu.P. Dotsenko, V.V. Naumov, A.N. Shmyryeva, P.S. Smertenko, Solar Energy Materials & Solar Cells **76**, 4, 613 (2003)
24. P. Smertenko, L. Fenenko, L. Brehmer and S. Schrader, Advantages in Colloid and Interface Science **116**, 255 (2005)
25. T.Ya. Gorbach, R.Yu. Holiney, L.A. Matveeva, P.S. Smertenko, S.V. Svechnikov, E.F. Venger, R. Ciach, M. Faryna, Thin Solid Films **336**, 63 (1998).
26. M. Lampert, P. Mark, *Current Injection in Solids*, Academic Press, New York & London, 1970
27. A. N. Zyuganov, S. V. Svechnikov. *Contact-injection phenomena in semiconductors* (Naukova dumka, Kyiv, 1981)
28. S. M. Sze, *Physics of semiconductor devices* (New York, Chichester, Brisbane, Toronto, Singapore, 1981)
29. M. Pope, Ch.E. Swenberg, *Electronic Processes in Organic Crystals and Polymers*, second edition (New York, Oxford, Oxford University Press, 1999)
30. N.A. Davidenko, L.I. Fenenko, A.A. Ishtchenko, M. Kuzma, P.S. Smertenko, S.V. Svechnikov, Synthetic Metals **122**, 173 (2001)

DOI: 10.1051/epjconf/201713301004

Evolution of MBE HgCdTe defect structure studied with ion milling method

Malgorzata Pociask-Bialy *

Faculty of Mathematics and Natural Sciences, University of Rzeszow
S.Pigonia 1, 35-959 Rzeszow, Poland

Abstract. In this paper, is shown how ion milling can assist in assessing the defect structure of MCT by revealing the residual doping, and establishing the minimum level of donor concentration N_{md}, which is needed for obtaining n-regions with a reproducible n value. For this purpose, a study of the electrical properties of ion-milled LWIR n-type MCT films, un-doped and doped with indium with the concentration $N_{In} = 5 \times 10^{14}–10^{17}$ cm^{-3} is proposed .

1 Introduction

MCT still is holding its position as the fundamental material for midwave infrared and long- (LWIR) photodetectors. Processing MCT with low-energy ions, about 0,5 keV, causes the formation of a source of mercury interstitials Hg_I, which get injected into the bulk of the material due to a very high surface concentration [1,2]. Ion milling (IM) propagating into the crystal, the interstitials annihilate the mercury vacancies and react with acceptors to form donor centres. In p-type MCT, this converts a surface layer to n-type and creates a p–n junction. In n-type MCT these processes modify the electrical properties of the material [3].

Of the number of techniques used to characterize the quality of MCT, ion milling (IM) is emerging as a unique means to reveal electrically active defects and complexes. It is known that IM is capable of strongly affecting electrical properties of MCT, up to conductivity type conversion in *p*–type material [4]. It appears, that strongly non-equilibrium processes which take place under IM, when material is oversaturated with interstitial mercury atoms Hg_I generated near the surface [1-2], lead to the appearance of specific defect complexes, which may not form under other conditions. These complexes may also comprise intrinsic point and extended defects that normally do not show their presence due to their electrical neutrality or compensation [4,10].

By carefully measuring the parameters of the crystal before and after the milling, and following the disintegration of defects with time after IM ('relaxation'), one can detect and identify these defects.

2 Experiment

In this paper we report on applying IM for the study of defects in LWIR MCT films grown by MBE on GaAs substrates with ZnTe/CdTe buffer layers. The results obtained on the MBE films grown on GaAs are compared to those acquired on the MBE films grown on CdZnTe substrates, and on MCT bulk crystals and films grown by isothermal vapor (ISOVPE) phase epitaxy [8].

We have studied nominally un-doped MBE films with composition x=0.22 (LWIR) and x=0.30 (MWIR), and those *in situ* doped with an acceptor (As with concentration $10^{15}–10^{16}$ cm^{-3}) or a donor (In with concentration $5 \cdot 10^{14}–10^{17}$ cm^{-3}), both as-grown and after various kinds of post-growth annealing.
The thickness of the MBE films was 7 to 11 μm, and they were protected by ~ 0.3 μm–thick (top) and ~ 1.0 μm–thick (bottom) graded–gap films with x increasing up to ~ 0.5 near the both interfaces.
In a typical structure, a $Hg_{1-x}Cd_xTe$ (x=0.22 for a LWIR structure) 'absorber' film with a thickness d ~ 8–10 μm was protected by ~ 0.3 μm–thick (top) and ~ 1.0 μm–thick (bottom) graded–gap films with x increasing up to ~ 0.5 near the both interfaces. LWIR films (all investigated samples), both un-doped and *in situ* In-doped, were grown at a substrate temperature T_{sub} ~185 ^{0}C.
Ion milling was performed using an IB-3 (EIKO, Japan) etching system with Ar^+ ion energy 500 eV, current density 0.2 mA/cm^2 and milling time 10 min. The temperature of the samples during the milling was kept at ~293 K by means of cooling of the sample holder with water. The relaxation of the electrical properties of

* Corresponding author: pociask@ur.edu.pl

DOI: 10.1051/epjconf/201713301004

the samples was studied by measuring repetitively the Hall coefficient R_H and σ in magnetic field B of 0.01 up to 1.5 T. The measurements were performed on square-shaped van der Pauw structures. During the measurements the samples were kept at liquid N_2 temperature. The $R_H(B)$ and $\sigma(B)$ dependences were analyzed using Discrete Mobility Spectrum Analysis (DMSA), as described elsewhere [4,9].
First measurements were performed straight after the milling, with the allowance for the time required for mounting samples on the holders and cooling them down to LN_2 temperature. This took about 8-15 min for the electrical measurements, and about one hour for the PL experiments. After the measurements, the samples were allowed to warm up to the room temperature, and were stored in air before the next measurement, again performed at the low temperature.

3 Results and discussion

Using IM, a unique neutral defect, which forms at the stage of the growth, and which appears to be specific to our MBE films, was revealed.
Straight after IM the electron concentration at T=77 K $n_{77}(i)$ in all the MBE films (~10^{17} cm^{-3} for LWIR films) greatly exceeded that in the bulk and ISOVPE samples (see Fig. 1).

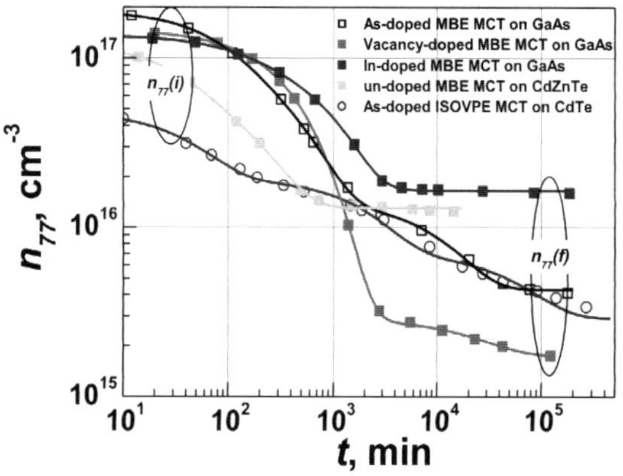

Fig. 1. An example of long-time relaxation of electron concentration after IM in some of the $Hg_{1-x}Cd_xTe$ (x≈0.22 for a LWIR) films studied. Points are experimental data and curves are multi-order exponential fits.
□ As-doped MBE MCT on GaAs, ■ Vacancy-doped MBE MCT on GaAs, ■ In-doped MBE MCT on GaAs, ■ un-doped MBE MCT on CdZnTe, ○ As-doped ISOVPE on CdTe.

The modification front, which forms under the milling, is very sharp, as the Hg_I migration does not affect the remaining 'core' of the crystal [1]. The properties of the 'core' are determined by the growth or post-growth treatment conditions. As to the modified n-layer, the electron concentration n there is usually determined by the total concentration of stable residual donors N_{SD} (both native N_{ID} and extrinsic N_{ED}) and the formed donor centres $D\cdot$ [5] When the milling stops, the centres gradually disintegrate. As a result of this relaxation process, n decreases, so within 10^3–10^5 min, depending on the chemical nature of acceptors, it stabilizes at the N_{SD} level [6]. In this way, in a device structure formed by ion milling in nominally un-doped MCT the electrical properties of the n-region are defined by residual donors.

For the $n_{77}(i)$ the following expression [10,11] can be write:

$$n_{77}(i) = N_{SD} + D^{\cdot} + N_N\,,$$

where N_N is the concentration of neutral defects, which get electrically activated through the interaction with Hg_I. These defects, as it was underlined earlier, form during the growth and may represent complex native defects, or initially electrically neutral impurity [9]. Cause of presence of the excessive tellurium in the films, possibly in a form of defect clusters [12], these defects may be Te-related.
After the relaxation, with the donor centres and electrically activated defects disintegrated, we have that

$$n_{77}(f) = N_{SD}.$$

In the In-doped LWIR films under study: $N_{SD} = N_{In} + N_{ID} + N_{ED}$ and it is clearly visible on Figure 1 that after ~5×10^3 min of ageing the n_{77} in the In-doped LWIR films (sample ■) with $N_{In} \geq 10^{16}$ cm^{-3} stabilised at the level of the concentration of the introduced dopant. As we see $n_{77}(f) = N_{In}$ for the case of sample ■. In nominally un-doped sample ■ the n_{77} stabilised at ~2×10^{15} cm^{-3}. For these samples it appears that $N_{ID} + N_{ED} > N_{In}$ and $n_{77}(f) \approx N_{ID} + N_{ED}$, so the $n_{77}(f)$ reflects the residual donor doping level. Given that the $n_{77}(f) \approx 2\times10^{15}$ cm^{-3} we can conclude that the as-grown films with $n_{77}(0) < 10^{15}$ cm^{-3} must been, in fact, compensated. N_{SD} typically lies in the range 10^{14}–10^{15} cm^{-3} and Its typical for molecular beam epitaxy (MBE) growth technique, conditions, as well as on the chemical purity of the MCT components used. To avoid the uncertainty related to repeatability of N_{SD}, intentional donor doping of the material can be used only for the one presented sample. It is believed that ion milling of donor-doped MCT should result in the fabrication of n-type material, whose electron concentration reflects the doping level [7,8]. In reality, however, in low-doped MCT residual donors (N_{SD}) may override intentionally introduced dopants [9]. Thus, to control the electron concentration in low-doped n-regions formed with ion milling, precise knowledge of the residual doping level is desirable.

The $n_{77}(i)$ in the MBE films depended only on x (i.e., growth conditions), and neither on substrate type, nor on the type and concentration of dopants, nor on the terms of post-growth annealing (Fig. 1).

Conclusions

Thus, it was concluded that in our MBE films IM led to the formation of a donor center, which resulted from the interaction of Hg_I with a certain defect that had been electrically neutral before the milling. The analysis showed that the most probable candidate was an initially neutral impurity (e.g., oxygen), or a defect complex involving excessive tellurium atoms. The exact nature of

DOI: 10.1051/epjconf/201713301004

the defect and its relation to the MBE growth conditions will be discussed in details.

Using IM, we also established donor and acceptor background doping level in the MBE films and made estimates of their electrical compensation. The electron concentration as measured after the relaxation of defects formed under the milling, $n_{77}(f)$, appears to be similar in all the HgCdTe samples (except for those strongly doped with In), ~(1-4)×10^{15} cm^{-3}. We relate this feature to the MBE growth conditions that allow for a variety of Te-related defects.

References

1. V.V. Bogoboyashchyy and I.I. Izhnin, Russ. Phys. J. **43,** 627–36 (2000)
2. D. Shaw and P. Capper, J. Mater. Sci.: Mater. Electron. **11,** 169–177 (2000)
3. K.D. Mynbaev and V.I. Ivanov-Omskii, Semiconductors **37,** 1127–50 (2003)
4. I.I. Izhnin, S.A. Dvoretsky, N.N. Mikhailov, Yu.G. Sidorov, V.S. Varavin, K.D. Mynbaev, and M. Pociask, Appl. Phys. Lett. **91,** 132106 (2007)
5. I.I. Izhnin, V.V. Bogoboyashchyy and F. F. Sizov, Proc. SPIE **5881** 5881OU1–11 (2005)
6. O. Gravrand, E. De Borniol, S. Bisotto, L. Mollard and G. Destefanis, J. Electron. Mater. **36,** 981–987 (2007)
7. K. R. Kurbanov and V.V. Bogoboyashchii, Condens. Media Interface **1**, 245–249 (1999)
8. V.S. Varavin, V.V. Vasiliev, S.A. Dvoretsky, N.N. Mikhailov, V.N. Ovsyuk, Yu.G. Sidorov, A.O. Suslyakov, M.V. Yakushev and A.L. Aseev, Opto–Electronics Review **11**, 99–111 (2003)
9. M. Pociask, I.I. Izhnin, S.A. Dvoretsky, N.N. Mikhailov, Yu.G. Sidorov, V.S. Varavin, K.D. Mynbaev and E. Sheregii, Semicond. Sci. Technol. **23**, 095001, 5pp (2008)
10. I.I. Izhnin, S.A. Dvoretsky, K.D. Mynbaev, N.N. Mikhailov, Yu.G. Sidorov, V.S. Varavin, R. Jakiela, M. Pociask, and G. Savitsky, *phys. stat. sol. (c)* **7**, 6 1618-1620 (2010)
11. K.D. Mynbaev, N.L. Bazhenov, M.V. Yakushev, D.V. Marin, V.S. Varavin, Yu.G. Sidorov, S.A. Dvoretsky, Tech. Phys. Lett. **40**, 708 (2014)
12. S. N. Yakunin and N. N. Dremova, JETP Letters **87,** 494–497 (2008)

Influence of infrared radiation on the electrical characteristics of the surface-barrier nanostructures based on MBE HgCdTe

Malgorzata Pociask-Bialy[1,*], *Ihor* Izhnin[2,3], *Alexander* Voitsekhovskii[3,4], *Sergey* Nesmelov[3,4], and *Stanislav* Dzyadukh[3,4]

[1]Faculty of Mathematics and Natural Sciences, University of Rzeszow, 35-310 Rzeszow, Poland
[2]Scientific Research Company "Carat", 79031 Lviv, Ukraine
[3]National Research Tomsk State University, 634050 Tomsk, Russia
[4]Siberian Physical-Technical Institute TSU, 634050 Tomsk, Russia

Abstract. Impact of illumination on the admittance of the MIS structures based on MBE $Hg_{1-x}Cd_xTe$ with graded-gap layers and single quantum wells was investigated. It is shown that for HgCdTe-based nanostructures the illumination greatly affects the capacitance and conductance dependencies. The capacitance-voltage characteristics exhibit a low-frequency behavior, which is associated with a decrease in the differential resistance of the space charge region. Especially informative illumination exposure is in the study of deep traps in *n*-HgCdTe (x=0.21-0.23) without graded-gap layer. Illumination leads to the low-frequency behavior of capacitance-voltage characteristics of MIS structures based on *p*-HgCdTe with HgTe single quantum well in the active region, and maximums in the voltage dependences do not appear.

1 Introduction

Nanostructures based on MBE $Hg_{1-x}Cd_xTe$ are promise for a new generation of devices for optoelectronics and nanophotonics [1]. Interest in MIS structures based on the HgCdTe is need to passivate the developed devices surface. The usually used method for studying of the MIS structure is a measurement of admittance in various conditions. Additional information on the physical processes in structures can give admittance measurement at illumination [2, 3].

This work presents the first results of investigation of the admittance at illumination impact for MIS structures based on HgCdTe with nanoinclusions in the active region (graded-gap layers, single quantum wells).

2 Samples and methods

MIS structures have been prepared on the basis of *n*-$Hg_{0.78}Cd_{0.22}Te$ grown by MBE on the Si (013) substrates. During growth of the heterostructures, graded-gap layers with high CdTe content were created on both sides of the working layer (figure 1). Before the formation of the insulator coatings, the conductivity and concentration and mobility of majority charge carriers were investigated by Hall measurements at 78 K. We investigated the MIS structures based on *n*-$Hg_{0.78}Cd_{0.22}Te$ / Al_2O_3 with near-surface graded-gap layer (structure No.1) and without graded-gap layer (structure No.2). The average electron concentration in the epitaxial film was 5.4×10^{14} cm^{-3} and electron mobility was 34000 cm^2/(V×c). For the structure No. 2, the near-surface graded-gap layer was removed by etching in the Br_2-HBr solution and then, a Al_2O_3 insulator was deposited used by plasma enhanced atomic layer deposition [4]. For the structure No. 1, the insulator coating was formed over the graded-gap layer. The study of these samples allows to investigate the influence of graded-gap layers on the properties of MIS structures.

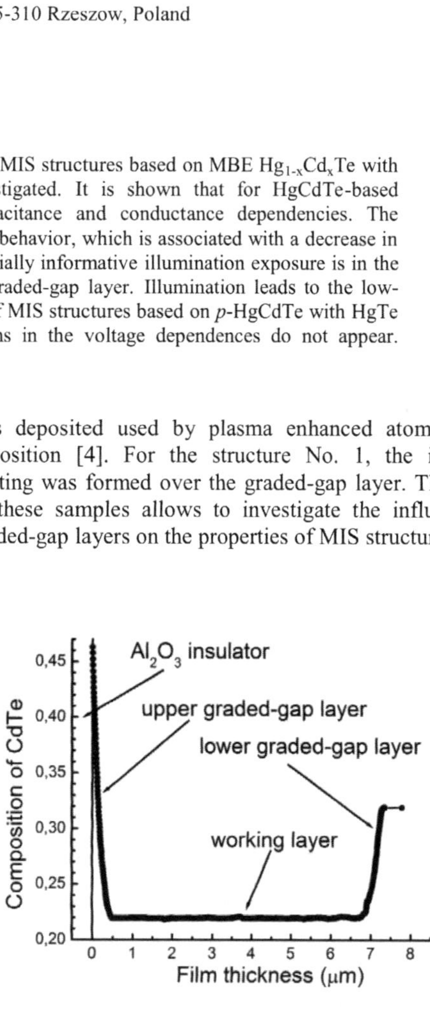

Fig. 1. Distribution of the composition over the thickness of the structure No. 1 based on MBE $Hg_{1-x}Cd_xTe$ measured with an automatic ellipsometer during the growth process.

The samples with HgTe single quantum well (SQW) were fabricated by MBE method based on *p*-HgCdTe on the GaAs (013) substrates at the Institute of Semiconductor Physics of SB RAS (Novosibirsk, Russia). The SQW thickness was 5.6 nm. $Hg_{0.35}Cd_{0.65}Te$ barrier

* Corresponding author: i.izhnin@carat.electron.ua

layers had a thickness of 35 nm. A double-layer insulator SiO_2/Si_3N_4 was deposited over the CdTe 40 nm layer. Figure 2 shows the basic layout of the layers in the structure with SQW.

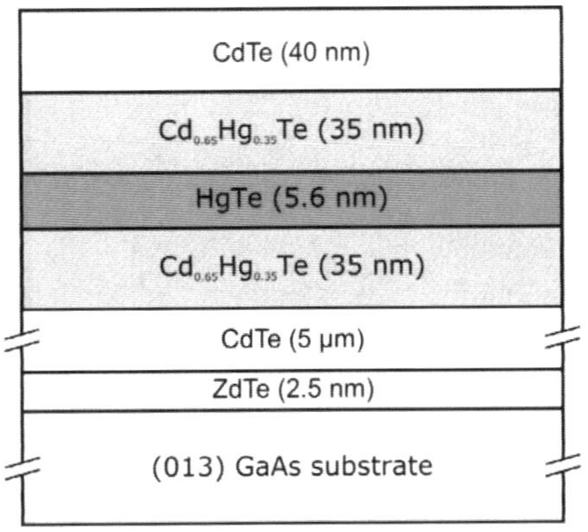

Fig. 2. Scheme of the layers position in structure with SQW.

Measurements were performed with an automated setup for spectroscopy of the nanoheterostructure admittance based on a cryostat Janis and an immittance meter Agilent E4980A. In measurements, the voltage change from negative values to positive ones. Electrophysical characteristics were measured in the dark mode and at illumination through the substrate at light wavelength of 0.94 microns.

3 Results and discussion

3.1 MIS structures based on MBE *n*-HgCdTe with graded-gap layers

Figure 3 shows the capacitance and conductance versus voltage for MIS structures based on n-$Hg_{0.78}Cd_{0.22}Te$ with (structure No.1) and without (structure No. 2) a graded-gap layer at temperature 15 K for dark mode and at illumination.

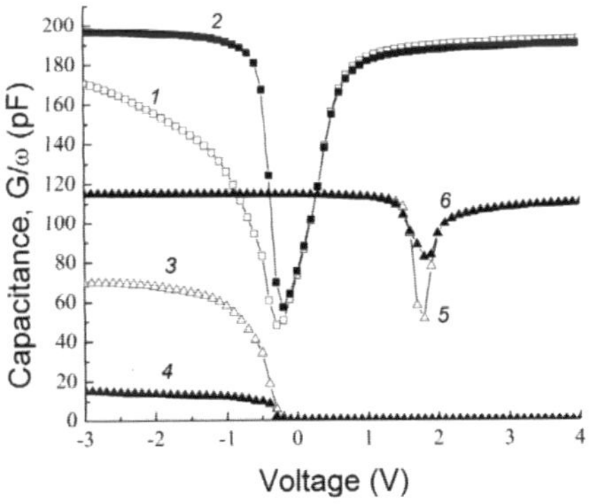

Fig. 3. Voltage dependencies of capacitance (curves 1, 2, 5, 6) and conductance at frequency 20 kHz (curves 3, 4) for MIS structures No. 1 (curves 1-4) and No.2 (curves 5, 6) measured at 15 K in dark mode (curves 1, 3, 5) and at illumination (curves 2, 4, 6). (где ось проводимости?)

Figure 4 shows the values of the equivalent circuit elements of the MIS structure in strong inversion determined from experimental data by the method described in [5]. From Figure 4 can be seen that the capacitance of the inversion layer under illumination increases, due to the increase in the concentration of minority carriers of charge. The illumination reduces the differential resistance of the space charge region (SCR) due to the appearance of an additional mechanism of generation of minority carriers (photogeneration). This reduces the time constant recharging of the inversion layer. The capacitance of the SCR under illumination is somewhat increased, due to the decrease in the maximum thickness of the SCR due to the increase in the concentration of minority carriers in the inversion layer.

From the analysis of Fig. 4, it can be concluded that a illumination strongly changes the capacitance-voltage (CV) characteristics of MIS structures based on $Hg_{0.78}Cd_{0.22}Te$ with a near-surface graded-gap layer. CV curves at illumination demonstrates a more low-frequency behavior than in the dark mode, and when illuminated capacitance increases in the minimum CV curves. At low frequencies, the illumination decreases the conductance of MIS structures based on HgCdTe with graded gap layer, and at high frequencies - to increase conductance. For structure No.2 without the graded-gap layer the impact of illumination is less pronounced, as the CV curves exhibits low-frequency behavior of in the dark mode.

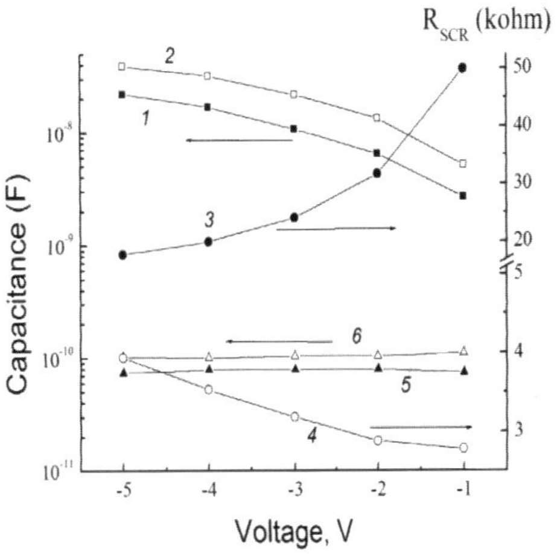

Fig. 4. Voltage dependencies of inversion layer capacitance (curves 1, 2), resistance of SCR R_{SCR} (curves 3, 4) and capacitance of SCR (curves 5, 6) for MIS structure No. 1 measured at temperature of 15 K at frequency range of 20-50 kHz in dark mode (curves 1, 3, 5) and at illumination (curves 2, 4, 6).

In order to calculate the ideal CV-characteristics at illumination, the Poisson equation for a given distribution of CdTe in the near-surface graded-gap layer was solved numerically [6]. In the calculations, we assumed local electroneutrality; the equation was solved taking into

* Corresponding author: i.izhnin@carat.electron.ua

account the changes in the near-surface graded-gap layer permittivity, band gap, and intrinsic carrier concentration. Account of illumination conducted by adding to the equilibrium concentration of mobile carriers in the expression for the density of the charge carrier concentration encountered during photogeneration.

Figure 5 shows the dependence of semiconductor capacity in the minimum low-frequency CV characteristics calculated at different temperatures for a structure with graded gap layer and without the graded-gap layer, and the dependence of semiconductor capacity at high frequency in the strong inversion mode calculated for structures with graded-gap layer at a temperature of 77 K.

Convenient to use for the characterization of illumination intensity ratio p1/p0, where p1 - concentration of minority carriers under illumination, p0 - concentration of minority carriers in the dark mode in the semiconductor bulk [3]. It can be seen that for a structure without a graded-gap layer capacity increase in the minimum CV characteristics begins with small values p1/p0 (curve 5), for structures with graded gap layer at low values of p1/p0 capacity at the minimum low-frequency CV curves does not change, and the increase in capacity at the minimum CV characteristics occurs, starting with a certain value p1/p0, which increases with decreasing temperature. The reason for this is that the value of p0, calculated near the interface at the maximum content of CdTe is less than the mean effective concentrations of minority carriers in the SCR as the equilibrium dark concentration of minority carriers in the graded-gap layer increases with distance from semiconductor boundary.

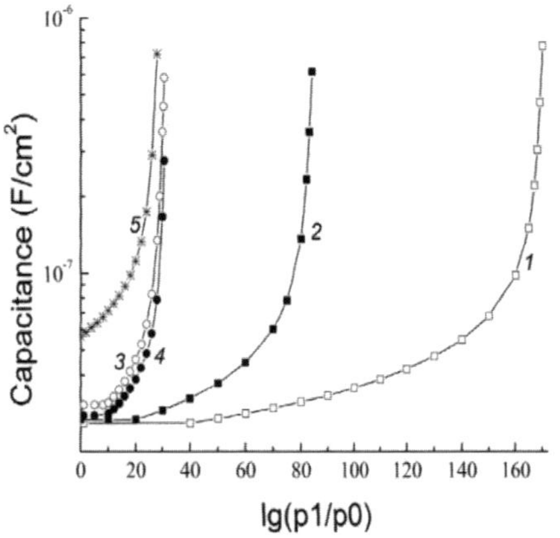

Fig. 5. Calculated dependencies of capacitance in minimum of low-frequency CV curves versus relation of minority carrier concentration at illumination (p1) to minority carrier concentration in dark mode for graded-gap n-$Hg_{0.78}Cd_{0.22}Te$ at electron concentration of $1.7*10^{15}$ cm^{-3} for various temperatures 15 (curve 1), 30 (curve 2), 77 K (curve 3), for n-$Hg_{0.78}Cd_{0.22}Te$ without graded-gap layer at 15 K (curve 5) and high-frequency capacitance in strong inversion mode for graded-gap n-$Hg_{0.78}Cd_{0.22}Te$ at 77 K (curve 4).

Investigation admittance under illumination for MIS structures based on *n*-HgCdTe in the strong inversion mode allows you to determine the value of the insulator capacitance, which is impossible for MIS structures based on *n*-HgCdTe in the accumulation mode due to the effects of degeneracy and non-parabolicity of conduction band.

3.2 MIS structures based on MBE *n*-HgCdTe with deep traps

Especially informative illumination exposure in the study of deep traps in *n*-HgCdTe without graded-gap layer. The appearance of deep traps in the CV characteristics under illumination only was observed (in dark mode, the traps levels are filled, and traps are not manifested in the measurements). If the deep traps appear in dark characteristics, the illumination typically reduces recharge time for deep traps which exhibit a low-frequency behavior at higher frequencies.

It is of interest that the special features associated with the capacity due to a deep level appear in the capacitance-voltage and volt-siemens characteristics of the MIS structures, for which the maximal photo-emf drop is observed under illumination in the strong inversion mode [7]. Without illumination, no special features were observed in the capacitance-voltage characteristics of some samples. Figure 6 shows the CV curves for the MIS HgCdTe/AOF structures without a graded-band-gap layer in the dark and on illumination. The capacitance-voltage characteristics for the illuminated MIS *n*-HgCdTe structures with different insulator coatings are found to differ from each other indicating that this deep level is rather due to the epitaxial-film properties than to the insulator-semiconductor interface. The special features in the voltage dependences of capacitance due to the deep level testify to the fact that the states located in the vicinity of the insulator-semiconductor interface (possible, at the interface between the natural oxide layer and epitaxial film) take part in the formation of capacitance properties of the structures. No special features were observed in the capacitance-voltage characteristics under illumination of samples produced from other wafers.

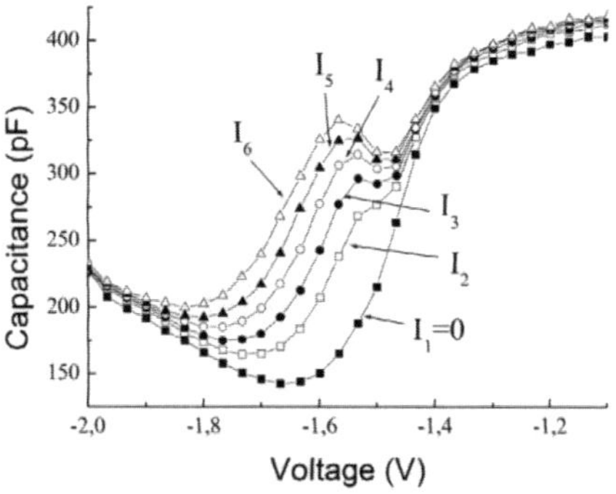

Fig. 6. CV curves for MIS structure based on n-$Hg_{0.77}Cd_{0.23}Te$ without graded-gap layer at 77 K at various intensity of illumination ($I_6>I_5>I_4>I_3>I_2>I_1=0$).

Estimations of the deep-level energy position made by the method of high-frequency capacitance-voltage characteristics using the dark capacitance-voltage

* Corresponding author: i.izhnin@carat.electron.ua

characteristic show that the observed deep level is 0.04–0.05 eV above the valence-band top. The level of a doubly ionized mercury vacancy, being an acceptor center, can have a similar energy position.

Thus, a levels can be present in the near-surface layer of the *n*-HgCdTe film. These levels determine the shape of the capacitance-voltage characteristics of the MIS structures under illumination.

3.3. MIS structures based on MBE *p*-HgCdTe with HgTe SQW

Figures 7 and 8 show dependencies of capacitance on voltage for MIS structure with SQW measured at a temperature of 8 K for various frequencies of test signal in dark mode and under illumination, respectively. It may be seen that HgCdTe has a *p*-type conductivity, and in the strong inversion mode at frequencies from (1-20) kHz non-monotonic change in capacitance is observed due to the recharging quantization levels [8]. At low frequencies CV characteristic have a low-frequency behavior and maxima become less pronounced than at frequencies of 1–20 kHz.

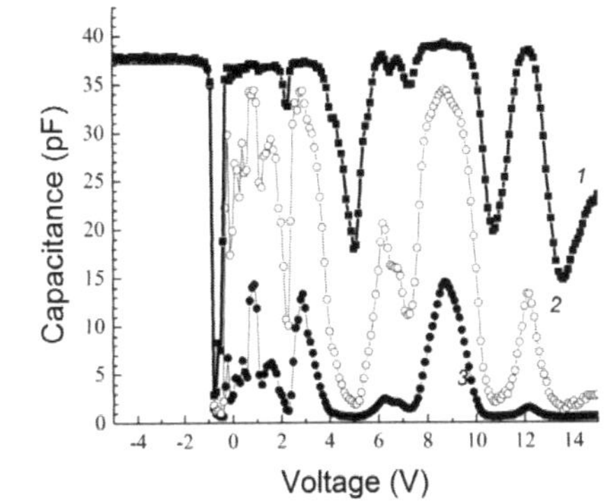

Fig.7. CV curves for MIS structures based on *p*-$Hg_{0.35}Cd_{0.65}Te$ with HgTe SQW measured at 8 K in dark mode at various frequencies, kHz: 1 – curve 1, 5 – curve 2, 20 – curve 3.

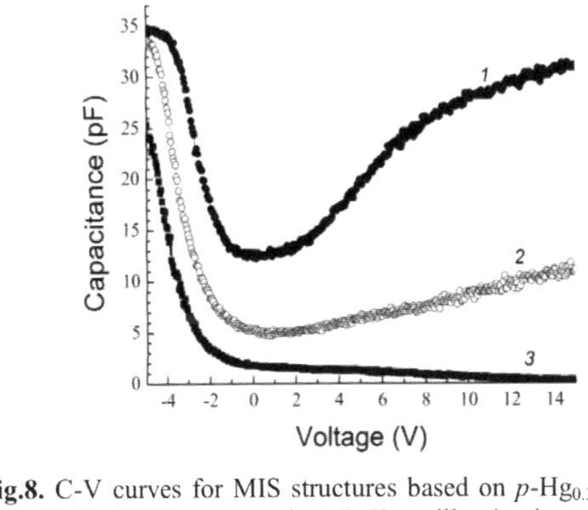

Fig.8. C-V curves for MIS structures based on *p*-$Hg_{0.35}Cd_{0.65}Te$ with HgTe SQW measured at 8 K at illumination at various frequencies, kHz: 1 – curve 1, 5 – curve 2, 20 – curve 3.

Under illumination maxima in the voltage dependences do not appear (Fig. 8). A possible cause is the photoexcitation of charge carriers from the quantum-well levels.

4 Conclusions

Thus, to study the effect of illumination on the electrical characteristics of MIS structures based on HgCdTe, including for cases presence of graded-gap layers and quantum wells in the active region.

CV measurements have shown that for structure with graded-gap layer under illumination reduces the time of formation of the inversion layer and an increase in capacitance value at the minimum low-frequency CV curves. Parallel differential conductance of MIS structure under illumination decreases at low frequencies, but increases at high frequencies. The changes under illumination element values equivalent circuit in strong inversion mode at different bias voltages were investigated. Low-frequency and high-frequency CV characteristics of MIS structures based on MBE *n*-$Hg_{0.78}Cd_{0.22}Te$ in the dark and under different illumination intensities were numerically simulated. For structures with graded gap layer at low illumination intensities, there is a reduction of the surface potential, but there are no changes in the capacitive characteristics.

It is noted that the deep levels manifest themselves in the capacitance-voltage characteristics upon illumination of the MIS structures based on *n*-$Hg_{0.78}Cd_{0.22}Te$ without graded-gap layer under study. The deep-level energy according to the admittance measurements is 0.04–0.05 eV of the valence-band top.

It is experimentally shown that for MIS structures based on MBE *p*-HgCdTe the presence of a SQW thickness of 5.6 nm can lead to the appearance of capacitance-voltage characteristics peaks in the strong-inversion mode. Under illumination maxima in the voltage dependences do not appear due to photoexcitation of charge carriers from the SQW levels.

This research was supported by the Russian Foundation for Basic Research (grant p_a No.16-42-700759).

References

1. A. Rogalski, Infrared detectors, New York: CRC Press, Taylor & Francis Group, 2011
2. V.G. Litovchenko, Surf. Sci., **1**, 3, 291 (1964)
3. J. Grosvalet, C. Lund, IEEE Trans. on Electron, Dev., ED-**14**, 11, 777 (1967)
4. R. Fu, J. Pattison, Opt. Engin., **51**, 104003 (2012)
5. A.V. Voitsekhovskii, S.N. Nesmelov, S.M. Dzyadukh, J. Electron. Mater., **45**, 2, 881 (2016)
6. A.V. Voitsekhovskii, S.N. Nesmelov, S.M. Dzyadukh, Thin Solid Films, **522C**, 261 (2012)
7. A.V. Voitsekhovskii, S.N. Nesmelov, S.M. Dzyadukh, Thin Solid Films, **551**, 92 (2014)
8. I.I. Izhnin, S.N. Nesmelov, S.M. Dzyadukh, A.V. Voitsekhovskii, D.I. Gorn, S.A. Dvoretsky, N.N. Mikhailov, Nanoscale Res. Lett., **11**, 1, 53 (2016)

* Corresponding author: i.izhnin@carat.electron.ua

Structural disordering studies of Cu_6PS_5I-based thin films deposited by magnetron sputtering

Ihor Studenyak[1,*], *Stefan* Rybak[1], *Andrii* Bendak[1], *Vitalii* Izai[1], *Pavlo* Guranich[1], *Peter* Kúš[2], and *Marian* Mikula[2]

[1]Faculty of Physics, Uzhhorod National University, 3 Narodna Sq., 88000 Uzhhorod, Ukraine
[2] Faculty of Mathematics, Physics and Informatics, Comenius University, Mlynska dolina, 84248 Bratislava, Slovakia

Abstract. Cu_6PS_5I-based thin films were deposited onto silicate glass substrates by magnetron sputtering. With Cu content increase, a red shift of the optical transmission spectra as well as increase of the total electric conductivity are observed. A typical Urbach bundle are revealed, the temperature behaviour of the Urbach absorption edge in thin films are explained by strong electron-phonon interaction. Temperature dependences of the absorption edge energy position and the Urbach energy for thin film are well described in Einstein model. The influence of different type of disordering on the Urbach tail is studied.

1 Introduction

Cu_6PS_5I superionic conductors belong to the compounds with argyrodite-type structure, characterized by high ionic conductivity and intrinsic structural disordering [1]. Due to the remarkable physical properties Cu_6PS_5I are promising materials for creating renewable energy sources, electrochemical and optical sensors. During the last decades the optical properties of Cu_6PS_5I crystals have been studied in detail [2]. However, the studies of the optical properties of Cu_6PS_5I thin films only begins. For the first time Cu_6PS_5I thin films were deposited and studied in Ref. [3]. It should be noted that the thin films based on superionic conductors can be applied to the production of supercapacitors of new generation [4].

Therefore, investigation of physical properties of Cu_6PS_5I-based thin films is of great interest for argyrodite-type superionic conductors. In the present paper, we study the electrical conductivity and temperature behaviour of optical absorption edge of Cu_6PS_5I-based thin films deposited by magnetron sputtering.

2 Experimental

For Cu_6PS_5I-based thin films sputtering we used the co-deposition technique from two tilted magnetrons, one equipped with Cu_6PS_5I target (pressed powder) and second with pure Cu target. The deposition was carried out at room temperature in Ar atmosphere. For target preparing the microcrystalline powder with the average grain size of 50 μm was obtained by grinding of the synthesized material in an agate mortar.

Energy-dispersive X-ray spectroscopy was used to ensure the thin films chemical composition. It is shown that thin films in interval from $Cu_{6.31}P_{1.10}S_{4.68}I_{0.91}$ to $Cu_{7.20}P_{0.83}S_{4.14}I_{0.83}$ are enriched by copper and phophorous (at low concentrations of Cu) atoms, and deficient in phosphorous (at high concentrations of Cu), sulphur and halogen atoms, besides with copper content increase the content of phosphorous, sulphur and halogen decrease.

Electrical conductivity was measured by impedancemeter at frequency 1 MHz. Optical transmission spectra were studied in the interval of temperatures 77–300 K by an MDR-3 grating monochromator, UTREX cryostat was used for low-temperature studies. Spectral dependences of the absorption coefficient and dispersion dependences of refractive index of thin films were calculated from the transmission spectra [5].

3 Results and discussion

Electrical studies have shown that the total electric conductivity of the thin films increase with increase of Cu atoms content. Thus, with Cu content increase in interval from $Cu_{6.31}P_{1.10}S_{4.68}I_{0.91}$ to $Cu_{7.20}P_{0.83}S_{4.14}I_{0.83}$ the electric conductivity increase from 0.053 S/m to 0.060 S/m, respectively. The high value of electrical conductivity in thin films under investigation make them the promising material for creation of solid state batteries and supercapacitors.

Futhermore, with Cu content increase, a red shift of the optical transmission spectra as well as absorption edge spectra is observed. Temperature variations of optical transmission in Cu_6PS_5I-based thin films are similar for all investigated samples (8 samples), but we will illustrate it for $Cu_{6.35}P_{1.12}S_{4.57}I_{0.95}$ thin film. Thus, interferential transmission spectra of $Cu_{6.35}P_{1.12}S_{4.57}I_{0.95}$ thin film at various temperatures within 77–300 K are shown in Fig. 1. With temperature increase, a red shift of both the short-wavelength part of the transmission spectrum and the interferential maxima is observed.

* Corresponding author: studenyak@dr.com

DOI: 10.1051/epjconf/201713302002

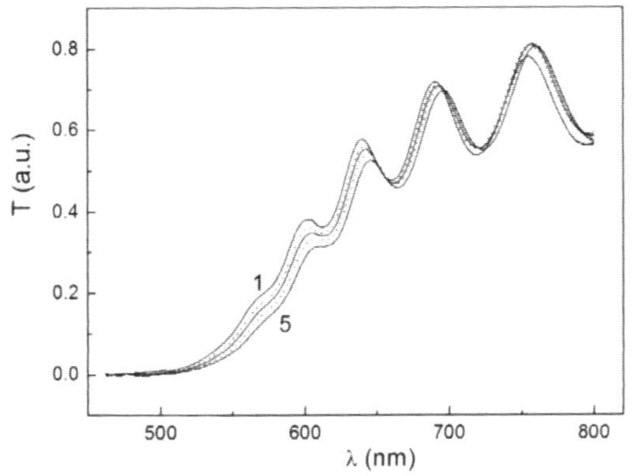

Fig. 1. Optical transmission spectra of $Cu_{6.35}P_{1.12}S_{4.57}I_{0.95}$ thin film at various temperatures: (1) 77, (2) 150, (3) 200, (4) 250 and (5) 300 K.

Figure 2 shows that the optical absorption edge spectra in the range of their exponential behaviour in $Cu_{6.35}P_{1.12}S_{4.57}I_{0.95}$ thin film are described by the Urbach rule [6]

$$\alpha(h\nu,T)=\alpha_o\cdot\exp\left[\frac{\sigma(h\nu-E_0)}{kT}\right]=\alpha_o\cdot\exp\left[\frac{h\nu-E_0}{E_U(T)}\right], \quad (1)$$

where E_U is the Urbach energy, σ is the absorption edge steepness parameter, α_0 and E_0 are the convergence point coordinates of the Urbach bundle. The coordinates of the Urbach bundle convergence point α_0 and E_0 for the $Cu_{6.35}P_{1.12}S_{4.57}I_{0.95}$ thin film are 1.13×10^6 cm^{-1} and 3.322 eV.

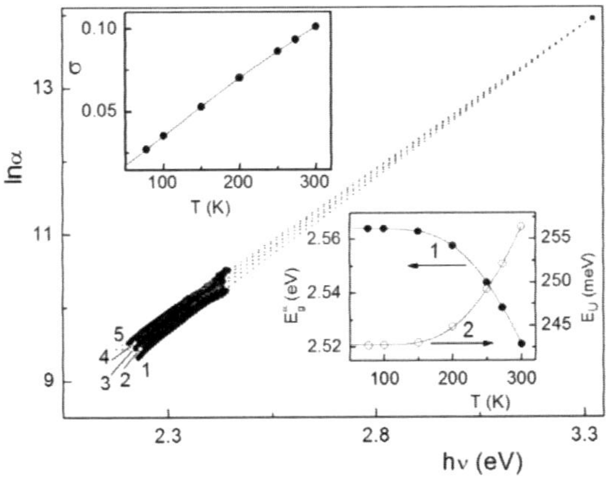

Fig. 2. Spectral dependences of the absorption coefficient of $Cu_{6.35}P_{1.12}S_{4.57}I_{0.95}$ thin film at various temperatures: (1) 77, (2) 150, (3) 200, (4) 250 and (5) 300 K. The insets show (a) the temperature dependence of the steepness parameter σ, (b) the temperature dependences of the absorption edge energy position E_g^α ($\alpha=5\times10^4$ cm^{-1}) (1) and Urbach energy E_U (2).

The temperature behaviour of the Urbach absorption edge in $Cu_{6.35}P_{1.12}S_{4.57}I_{0.95}$ thin film is explained by electron-phonon interaction (EPI) which is strong in the film under investigation. The EPI parameters are obtained from the temperature dependence of absorption edge steepness parameter (Fig.2) using the Mahr formula [7]

$$\sigma(T)=\sigma_0\cdot\left(\frac{2kT}{\hbar\omega_p}\right)\cdot\tanh\left(\frac{\hbar\omega_p}{2kT}\right), \quad (2)$$

where $\hbar\omega_p$ is the effective phonon energy in a one-oscillator model, describing the electron-phonon interaction (EPI), and σ_0 is a parameter related to the EPI constant g as $\sigma_0=(2/3)g^{-1}$ (parameters $\hbar\omega_p$ and σ_0 are given in Table 1). For the $Cu_{6.35}P_{1.12}S_{4.57}I_{0.95}$ thin film $\sigma_0<1$ that is the evidence for the strong EPI [8].

It should be noted that in the range of exponential behaviour of optical absorption for their. For spectral characterisation of optical absorption edge we used the energy position of an exponential absorption edge E_g^α values taken at $\alpha=5\times10^4$ cm^{-1} (Table 1). It should be noted that at T=300K E_g^α=2.521 eV, while E_U=256 meV. The temperature dependences of E_g^α and the Urbach energy E_U for $Cu_{6.35}P_{1.12}S_{4.57}I_{0.95}$ thin film (Fig.2) can be described in Einstein model by relations [9, 10]

$$E_g^\alpha(T)=E_g^\alpha(0)-S_g^\alpha k\theta_E\left[\frac{1}{\exp(\theta_E/T)-1}\right], \quad (3)$$

$$E_U(T)=(E_U)_0+(E_U)_1\left[\frac{1}{\exp(\theta_E/T)-1}\right], \quad (4)$$

where $E_g^\alpha(0)$ and S_g^α are the energy position of absorption edge at 0 K and a dimensionless constant, respectively; θ_E is the Einstein temperature, corresponding to the average frequency of phonon excitations of a system of non-coupled oscillators, $(E_U)_0$ and $(E_U)_1$ are constants. The obtained $E_g^\alpha(0)$, S_g^α, θ_E, $(E_U)_0$, and $(E_U)_1$ parameters for the thin film are given in Table 1.

An essential characteristic of the absorption edge spectra of the thin films under investigation is a lengthy Urbach tail which results in the high values of Urbach energy E_U (Table 1). In Ref. [11] it was shown that temperature and structural disordering affect Urbach absorption edge shape, i.e. the Urbach energy E_U is described by the equation

Table 1. Parameters of Urbach absorption edge and EPI for $Cu_{6.35}P_{1.12}S_{4.57}I_{0.95}$ thin film

α_0 (cm^{-1})	E_0 (eV)	σ_0	$\hbar\omega_p$ (meV)	θ_E (K)	$(E_U)_0$ (meV)	$(E_U)_1$ (meV)	$E_g^{\alpha}(0)$ (eV)	S_g^{α}
1.13×10^6	3.322	0.2	97.3	1129	243	575	2.564	18.6

$$E_U = (E_U)_T + (E_U)_X = (E_U)_T + (E_U)_{X,stat} + (E_U)_{X,dyn}, \quad (5)$$

where $(E_U)_T$ and $(E_U)_X$ are the contributions of temperature-related and structural disordering to E_U, respectively; $(E_U)_{X,stat}$ and $(E_U)_{X,dyn}$ are the contributions of static structural disordering and dynamic structural disordering to $(E_U)_X$, respectively. It is shown that the contribution of static structural disordering into the film Urbach energy equals 95%. Static structural disordering in $Cu_{6.35}P_{1.12}S_{4.57}I_{0.95}$ thin film may be additionally increase due to (1) the absence of long-range order in the atomic arrangement and chemical bond breakdown; (2) lower density of the atomic structure packing due to the presence of pores; (3) the transition from the three-dimensional bulk structure to the two-dimensional planar structure.

The dispersion dependences of the refractive index for the $Cu_{6.35}P_{1.12}S_{4.57}I_{0.95}$ thin film was obtained from the interference transmission spectra (Fig. 3). With temperature increase the nonlinear increase of refractive index in $Cu_{6.35}P_{1.12}S_{4.57}I_{0.95}$ thin film is revealed.

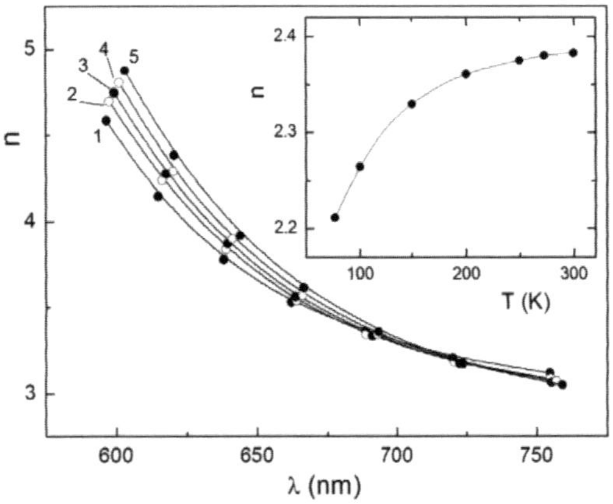

Fig. 3. Refractive index dispersions of $Cu_{6.35}P_{1.12}S_{4.57}I_{0.95}$ thin film at various temperatures: (1) 77, (2) 150, (3) 200, (4) 250 and (5) 300 K. The inset shows the temperature dependence of refractive index.

4 Conclusions

Cu_6PS_5I-based thin films were deposited onto silicate glass substrates by magnetron sputtering. Electrical conductivity of Cu_6PS_5I-based thin films was studied depends on chemical composition, with Cu content increase of the total electric conductivity was increased. Optical transmission spectra of $Cu_{6.35}P_{1.12}S_{4.57}I_{0.95}$ thin film was investigated in the temperature interval 77–300 K. With Cu content increase as well as with temperature increase, a red shift of the optical transmission spectra was revealed. Temperature variations of optical transmission in Cu_6PS_5I-based thin films are similar for all investigated samples, however it illustrated for $Cu_{6.35}P_{1.12}S_{4.57}I_{0.95}$ thin film. It is shown that the optical absorption edge spectra in the range of their exponential behaviour in $Cu_{6.35}P_{1.12}S_{4.57}I_{0.95}$ thin film are well described by the Urbach rule. Temperature dependences of the energy position of absorption edge, Urbach energy and refractive index of $Cu_{6.35}P_{1.12}S_{4.57}I_{0.95}$ thin film were analysed. The origin of lengthy Urbach tail and role of the different type of disordering were discussed.

Aknowledgments

This work has been supported by the Slovak Research and Development Agency under the contract numbers APVV-14-0173 and Operational Program Research and Development (projects ITMS:26240220027 and ITMS:26210120010). Vitalii Izai is grateful to the National Scholarship Programme of the Slovak Republic.

References

1. W.F. Kuhs, R. Nitsche, K. Scheunemann, Mat. Res. Bull. **11**, 1115-1124 (1976)
2. I.P. Studenyak, M. Kranjčec, M.V. Kurik, J. Phys. Chem. Solids **67**, 807-817 (2006)
3. I.P. Studenyak, M. Kranjčec, V.Yu. Izai, A.A. Chomolyak, M. Vorohta, V. Matolin, C. Cserhati, S. Kökényesi, Thin Solid Films **520**, 1729-1733 (2012)
4. A.L. Despotuli, A.V. Andreeva, B. Rambabu, Ionics **11**, 306-314 (2005)
5. R. Swanepoel, J. Phys. E: Sci. Instrum. **16**, 1214-1222 (1983)
6. F. Urbach, Phys. Rev. **92**, 1324-1326 (1953)
7. H. Sumi, A. Sumi, J. Phys. Soc. Japan **56**, 2211-2220 (1987)
8. M.V. Kurik, Phys. Stat. Sol. (a) **8**, 9-30 (1971)
9. M. Beaudoin, A.J.G. DeVries, S.R. Johnson, H. Laman, T. Tiedje, Appl. Phys. Lett. **70**, 3540-3542 (1997)
10. Z. Yang, K.P. Homewood, M.S. Finney, M.A. Harry, K.J. Reeson, J. Appl. Phys. **78**, 1958-1963 (1995)
11. G.D. Cody, T. Tiedje, B. Abeles, B. Brooks, Y. Goldstein, Phys. Rev. Lett. **47**, 1480-1483 (1981)

EPJ Web of Conferences **133**, 02003 (2017) DOI: 10.1051/epjconf/201713302003
IC SeNOB 2016

Electron magnetic resonance study of multiwalled carbon nanotubes and carbon nanohorns

Ireneusz Stefaniuk[1*], *Bogumil* Cieniek[1], and *Iwona* Rogalska[1]

[1] Faculty of Mathematics and Natural Sciences, University of Rzeszow, S. Pigonia 1, 35-310 Rzeszow, Poland

Abstract. Temperature Electron Magnetic Resonance (EMR) measurements of multiwalled carbon nanotubes (CNT) and carbon nanohorns (CNH) were performed in the temperature range 5.2 – 300 K. The asymmetric resonance lines with Dyson shape were observed. The g-value of multiwalled carbon nanotubes (CNT) and carbon nanohorns (CNH) was calculated. Analysis of the temperature dependences of the integral intensity of EMR spectra was carried out using the Curie-Weiss law and Curie temperature $\theta_{(CNT)}$ = 46,6 K and $\theta_{(CNT)}$ = 8.6 K were obtained. We showed that the localization processes observed in nanocarbon materials lead to local quantum transport of spins or carriers.

1 Introduction

Since their discovery by Iijima in 1991 [1] carbon nanotubes have attracted increasing attention from the material science community due to their exceptional electronic and mechanical properties [2].

Carbon nanotubes are promising candidates for a variety of applications due to their uncommon and versatile electrical and mechanical properties which recently generated an active area of research [3]. Possible technological applications include, among others, nanotransistors [4, 5], spin-polarized electron sources [6], hydrogen storage [7, 8], flat display panels [9] and magnetic inks.

Single nanotubes have been used to fabricate transistors [10] and show potential as the building block for new nanotechnologies. Potential applications of nanotubes include using them as an electron source in field emission displays [11], as an optical limiting material [12] and as an electromagnetic interference shielding material. However, there remains a need to characterize the purity of carbon nanotube powders and to develop a mechanism to isolate nanotubes.

Quasi-graphitic carbon nanoparticles (CNs) were found to show very interesting behaviour with respect to the localization of charge carriers. Due to their large specific surface area, CNs are very sensitive to the adsorption of various molecules. Their electronic properties strongly depend on their structure and interactions between the molecules adsorbed in pores and the pore walls (host–guest interactions).

The magnetism of carbon-based materials (i.e. graphene, graphite, carbon nanotubes, and fullerenes) has recently been the subject of intense research [13-16]. These materials exhibit exotic magnetism such as ferromagnetism above room temperature, which is believed to be induced by defects in the graphitic network.

Magnetic and electronic properties of multiwall carbon nanotubes has been presented in [17-19].

The main aim of this work it characterization properties of multiwalled carbon nanotubes (CNT) and carbon nanohorns (CNH) using Electron Magnetic Resonance (EMR). Serially produced material of CNT and CNH by Sigma-Aldrich was used as samples. Although measurements of similar nanoparticles are available in literature, but our target in future is to modify CNT and CNH to use it for example as sensors. This is the reason to measure EMR in wide temperature range from room down to liquid helium temperature.

2 Experiment

Multi walled carbon nanotubes (MWNTs, CNTs) were prepared by chemical vapor deposition (CVD). In chemical vapor deposition (CVD), a volatile precursor undergoes thermal decomposition at elevated temperatures to form a solid deposit on a substrate. The diameter of the tube ranges between 6-13nm and the average length is 10μm. The multiwalled nanotubes are stable in inert atmosphere up to a temperature of 3697°C. CVD production method is followed by HCl demineralization. Single-walled carbon nanohorns are made of graphene sheets with long cone-shaped tips with the diameters of 2 - 5 nm and the length of 40-50 nm. Thousands of carbon nanohorns form spherical aggregates of about 100 nm in diameter.

CNT multiwalled, more than 98% of carbon basis, and CNH as grown material was manufactured by Sigma-Aldrich.

The EMR spectra were investigated in a wide range of temperatures from 5 K up to 300 K using an EPR X-

* Corresponding author: istef@ur.edu.pl

band spectrometer (Bruker multifrequency and multiresonance FT-EPR ELEXSYS E580). Measurements at low temperatures were carried out using an Oxford Instruments continuous-flow helium cryostat (3.8 K to 400 K with the Helium Temperature Control System ER4112HV).

CNT and CNH samples were transferred into thin-walled capillary closed after filling. Samples have been exposed to atmospheric oxygen only during filling the capillary.

The EMR spectra for the samples described above are shown in Fig. 1. The shape of the EMR spectra is of a Dyson type [20], with a asymmetry parameter of the resonance line a/b=1.5 at 50 K, where a and b are amplitudes of the low field and high field halves of the spectrum, respectively.

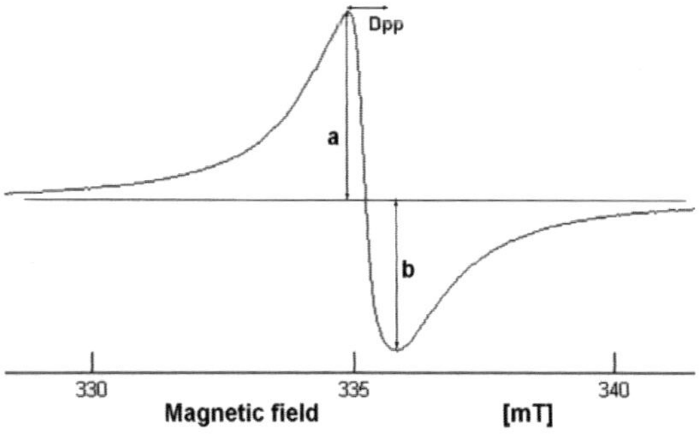

Fig. 1. Dysonian EMR signal of the CNH. The characteristic points and parameters are shown.

In Fig. 2 EMR spectra of CNT and CNH acquired at temperature 50 K are shown. In Fig. 3 and 4 the temperature dependences of EMR spectra are shown.

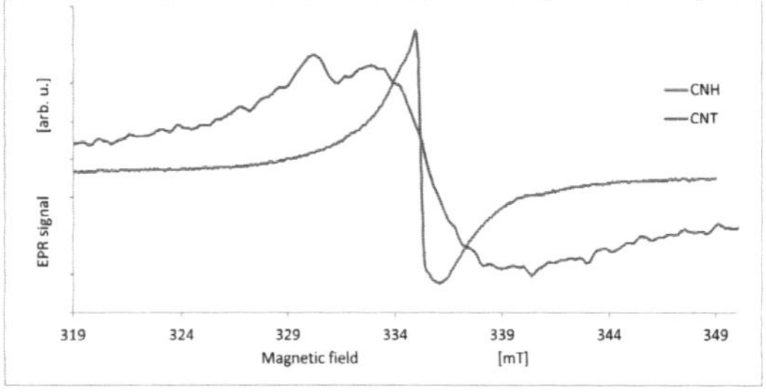

Fig. 2. The EMR spectra of the CNT and CNH at temperature 50K.

Two EMR signals (one broad and one narrow) were recorded at low temperatures for CNT sample. All spectra were asymmetric and could be well fitted to Dysonian shape. From these fits, the g value (representing position), Bpp (peak-to-peak linewidth) and integral intensity were calculated.

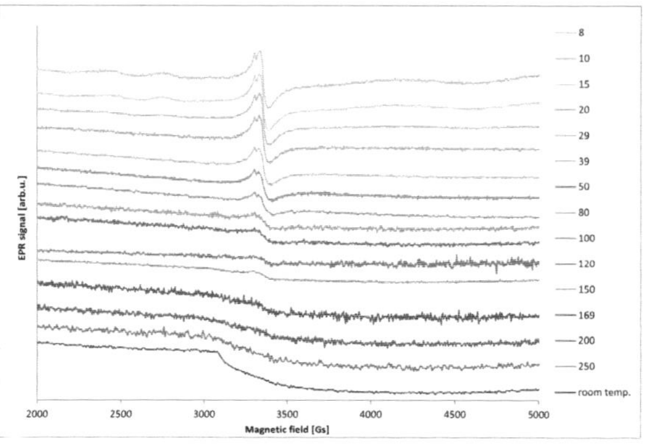

Fig. 3. The temperature dependence of EMR signal of the CNT.

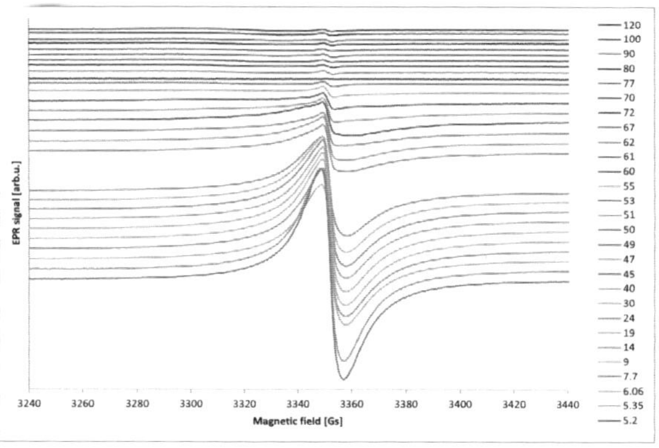

Fig. 4. The temperature dependence of EMR signal of the CNH.

At low temperatures, the EMR spectrum of CNH changes its shape to Lorenzian with only slight asymmetry (adduct of the Dysonian component).

3 Results and discussion

The selected EMR spectra as a function of temperature for the CNT and CNH sample are shown in Fig. 3 and 4, respectively. The temperature dependencies of the peak to peak linewidth Bpp and the estimated integral intensity

$$I_{int} = (A * B_{pp}^2) \quad (1)$$

of the signal (A = a + b) measured in the temperature interval 290 – 5.2 K.

In Fig. 5 and 6 linewidth B_{pp} for the CNH and CNT samples are presented respectively.

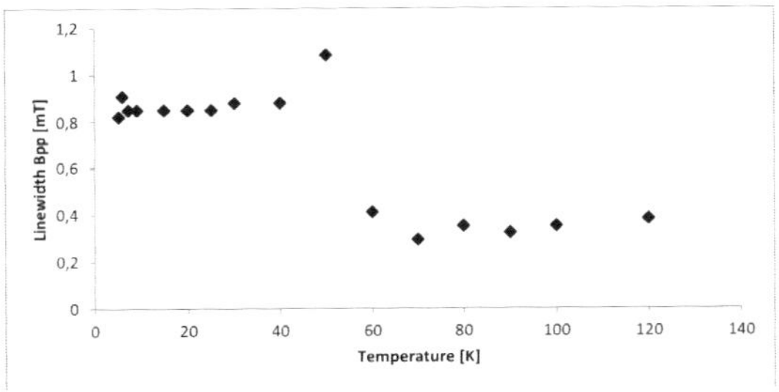

Fig. 5. The temperature dependence of linewidth B_{pp} of the CNH.

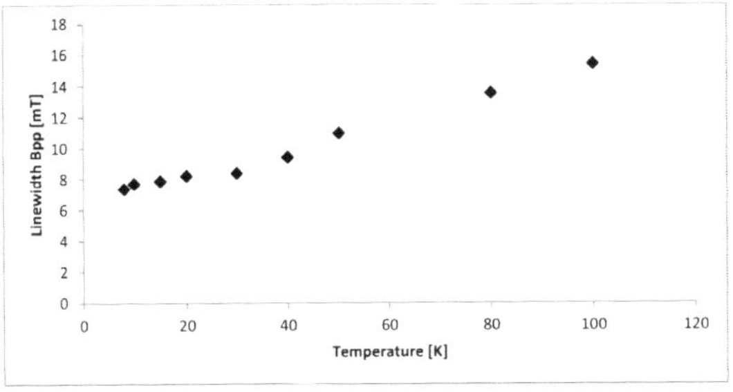

Fig. 6. The temperature dependence of line width B_{pp} of the CNT.

In Fig.7 and 8 integral intensity of EMR spectra obtained on eq. 1 for the CNH and CNT samples are presented respectively.

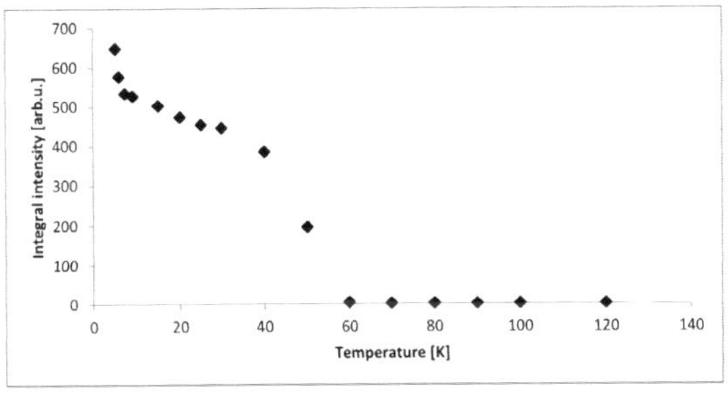

Fig. 7. The temperature dependence of integral intensity of the CNH.

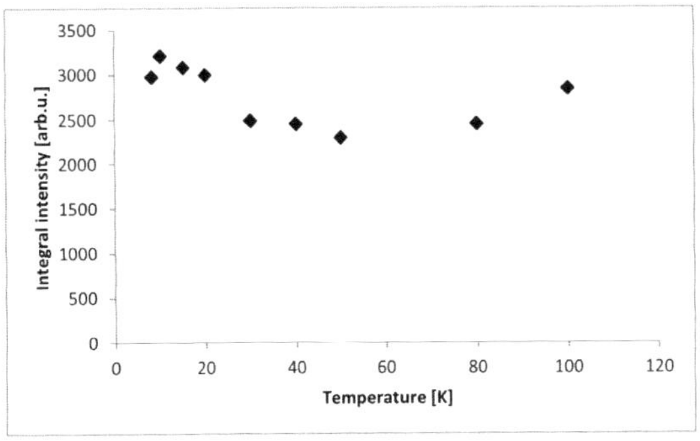

Fig. 8. The temperature dependence of integral intensity of the CNT.

The g-value was calculated using the magnetic field B_0, which is obtained from the central position of the EMR spectral line. The g-value of 2.01 – 2.02 is assigned to the interaction between the conducting electrons in the CNTs trapped at defects or magnetic ions site. The more the deviation from the freeelectron g-value, the more is the localization due to defects [21]. In CNH and CNT samples, EMR lines are seen with the g-value of 2.0027 – 1.999 at 100 K, 2.002 - 2.000 at 15 K respectively. Obtained value is very close to the free electron g-value (2.0023).

EMR is the perfect method to distinguish between the localized and conducting electrons and helps to determine the conducting properties of the carbon nanotubes and nanohorns. The conduction electrons can be considered as a Fermi gas. The alignment of magnetic moments with the external magnetic field is opposed not only by the thermal energy as for the isolated and independent electrons, but we also have to be sure that the Pauli principle is fulfilled. Therefore, the electron gas must obey to the Fermi-Dirac statistics [22]; consequently, the electrons that can be measured in a ESR experiment are those with energy near to the Fermi level. The others do not contribute. For this reason, the ESR spectra are usually less intense, regardless the number of electrons in the sample. The Fermi-Dirac statistic determines also the type of magnetism of these systems.

A solid that contains an ensemble of unpaired, non-interacting spins obeys to the Maxwell-Boltzmann statistic and its magnetic susceptivity obeys the Curie law [22, 23]

$$\chi = \frac{N\nu\mu^2}{3k_T T}\mu_0 = \frac{C}{T} \geq 0 \qquad (2)$$

where $\mu_0 = g^2\beta e\ 2S(S+1)$ and N_V is the number of magnetic species per unit volume. C is the Curie constant and is characteristic of each paramagnetic substance through μ. Usually, real solids do not obey exactly the Curie law; often the dependence of the magnetic susceptivity from the temperature has the following form

$$\chi = \frac{C}{T-\theta} \qquad (3)$$

this is the Curie-Weiss law and the parameter θ, the Weiss temperature, depends on the specific material and it is positive for materials that have a ferromagnetic behavior and negative for antiferromagnetic materials. The Curie-Weiss law comes as direct consequence of the dipolar magnetic interaction between the spin in the materials. Conductive systems have a magnetic suceptivity that is lower in module with respect to the systems with localized electrons and it doesn't depend on the temperature; in this case we have the Pauli paramagetism [22, 23].

A linear increase of χ–1 (T) at higher temperatures can be fitted to the Curie-Weiss law.

$$(\chi(T) - \chi(0))^{-1} = \frac{T-\theta}{C} \qquad (4)$$

where C is the Curie constant, θ is the paramagnetic Curie temperature, and $\chi(0)$ is a temperature independent term to account for the diamagnetic host and any Pauli paramagnetism contribution.

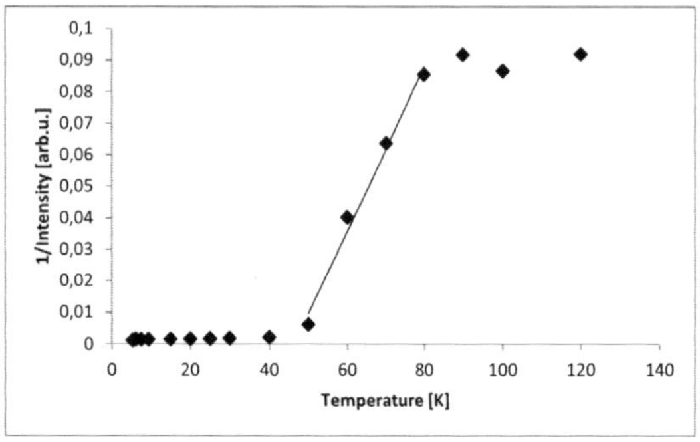

Fig. 9. Temperature dependence of the 1/(integral intensity) of CNH sample

Figure 9 displays the temperature dependence of the quantity $(\chi(T) - \chi(0))–1$ on CNH sample. The line is a linear extrapolation illustrating the ferromagnetic (positive) Curie-Weiss temperature. Fitting yields the following values $\theta = 46.6$ K and C = 384.6 of CNH. For CNT sample fitting yields the following values $\theta = 7.7$ K and C = 1234.3.

Therefore the EMR oscillations of CNT sample reflect jumplike displacements of impurities atom (i.e. oxygen) between walled carbon nanotubes. The amplitude of the oscillations depends on the number of electrons transferred from conduction band to localized state. Dysonian line characterized by two relaxation processes, relatively narrow Lorentzian line, characterized by one relaxation process. Such changes should modulate the conducting band as the result the conducting electrons give a broad EMR spectrum with fluctuating g-values.

The localized spin centers near defect states show Curie-like temperature dependence; by contrast, the EMR temperature dependence of conduction electron spins is close to that predicted by the Pauli magnetism. The comparison of EMR signal integral intensity at temperature dependence allows one to determine the relative contributions of both types of spin centres.

4. CONCLUSION

The asymmetric resonance lines with Dyson shape with an asymmetry parameter of the resonance line a/b=1.5 at 50 K were observed.

The g-value of multiwalled carbon nanotubes (CNT) and carbon nanohorns (CNH) was calculated. In CNH and CNT sample EMR lines are seen with the g-value of 2.0027 – 1.999 at 100 K, 2.002 - 2.000 at 15 K respectively.

Analysis of the temperature dependences of the integral intensity of EMR spectra was carried out using the Curie-Weiss law and Curie temperature $\theta_{(CNT)} = 46{,}6$ K and $\theta_{(CNT)} = 8.6$ K were obtained.

The localized spin centers near defect states show Curie-like temperature dependence; by contrast, the EMR temperature dependence of conduction electron spins is close to that predicted by the Pauli magnetism. The comparison of EMR signal integral intensity at temperature dependence allows one to determine the relative contributions of both types of spin centres.

References

1. S. Iijima, Nature **56**, 354 (1991)
2. M.S. Dresselhaus, G. Dresselhaus, P.C. Eklund, *Science of Fullerenes and Carbon Nanotubes* (Academic, New York, 1995)
3. M. Monthieoux, Carbon **40**, 1809 (2002)
4. K. Esfarjani, A.A. Farajian, F. Ebrahimi, Y. Kawazoe, Eur. Phys. J. D **16**, 353 (2001)
5. K. Tsukagoshi, I. Yagi, Y. Aoyagi, Appl. Phys. Lett. **85**, 1021 (2004)
6. S. Hao, G. Zhuo, J. Wu, W. Duan, B.L. Gu, Phys. Rev. B **69**, 113403 (2004)
7. A.C. Dillon, Nature **386**, 377 (1997)
8. C. Liu, Y.Y. Fan, M. Liu, H.T. Cong, H.M. Cheng, M.S. Dresselhaus, Science **286**, 1127 (1999)
9. W.A. Heer, A. Chatelain, D. Ugarte, Science **270**, 1179 (1996)
10. S.J. Tans, A.R.M. Verschueren, C. Dekker, Nature **393**, 49 (1998)
11. W.B. Choi, D.S. Chung, J.H. Kang, H.Y. Kim, Y.W. Jin, I.T. Han, Y.H. Lee, J.E. Jung, N.S. Lee, G.S. Park, J.M. Kim, Appl. Phys. Lett. **75**, 3129 (1999)
12. S.R. Mishra, H.S. Rawat, S.C. Mehendale, K.C. Rustagi, A.K. Sood, R. Bandyopadhyay, A. Govindaraj, C.N.R. Rao, Chem. Phys. Lett. **317**, 510 (2000)
13. O.V. Yazyev, Reports on Progress in Physics **73**, 056501 (2010)
14. W. Kempiński, S. Łoś, M. Kempiński, D. Markowski, Beilstein J. Nanotechnol. **5**, 1760 (2014)
15. W. Kempiński, M. Kempiński, D. Markowski, S. Łoś, Nukleonika **58**, 3, 371 (2013)
16. M.A. Augustyniak-Jabłokow, Y.V. Yablokov, B. B. Andrzejewski, W. Kempiński, S. Łoś, K. Tadyszak, M.Y. Yablokov, V.A. Zhikharev. Phys Chem Minerals **37**, 237 (2010)
17. V. Likodimos, S. Glenis, N. Guskos, C.L. Lin, Physical Review B **68**, 045417 (2003)
18. S. Glenis, V. Likodimos, N. Guskos, C.L. Lin, Journal of Magnetism and Magnetic Materials **272-276**, 1660 (2004)
19. V. Likodimos, S. Glenis, N. Guskos, C.L. Lin, Physical Review B **76**, 075420 (2007)
20. F.J Dyson., Physical Review. **98**, 349 (1955)
21. D.K. Singh, P.K. Iyer, P.K. Giri, J. Nanosci. Nanotechnol. **9**, 1 (2009)
22. N.W Ashcroft, N.D Mermin, *Solid State Physics* (Saunders College, Philadelphia, 1976)
23. A.F Orchard, *Magnetochemistry* (Oxford University Press, Oxford, 2003)

DOI: 10.1051/epjconf/201713302004

Comparison of Se and Te clusters produced by ion bombardment

Małgorzata Trzyna

Center for Microelectronics and Nanotechnology, Department of Mathematics and Natural Sciences, Rzeszow University, Poland

Abstract. Nanostructures based on tellurium and selenium are materials used as components for the manufacturing topological insulators. Therefore it is crucial to precisely characterize these materials. In this work the emission of selenium and tellurium cluster ions, sputtered by Bi^+ primary ion guns, was investigated by using Time-of-Flight Secondary Ion Mass Spectrometry (TOF SIMS). It has been found that Bi_xTe_x and Bi_xSe_x clusters appear in addition to Se_x and Te_x clusters in the mass range up to ~ 1300 m/z. Local maxima or minima (magic numbers) are observed in the ion intensity versus a number of atoms per cluster for both positive and negative ions spectra for all types of clusters and primary ions used. These extrema can be attributed to different yield and stability of certain clusters but also to fragmentation of high-mass clusters.

Introduction

Recently, the subject of topological insulators have attracted great attention in solid state physics. Most popular topological structures are based on elements of groups IV-VI and III-VI. Understanding properties of these compounds is crucial for their further development [1,2].
One of them is clusters formation process on the surface which influences the properties of the whole material. Cluster ions are frequently observed to be emitted from a variety of samples exposed to ion beam bombardment. The formation mechanism of these cluster ions is important if they are to be utilized to determine the structure of solid surfaces [3].
The clusters are made by different methods, however the most recent studies focus on the clusters obtained during material exposure to different primary ions. We have used the TOF SIMS as an analytical tool to measure species and clusters. Studies of the whole secondary ion spectrum allow to understand in greater depth how secondary ions are formed, which is of exceptional importance to the whole field of secondary ion mass spectrometry [4]. The purpose of this paper is to investigate the formation of Se and Te cluster ions using Bi^+ primary ions.

Materials And Methods

Two materials were chosen for the investigation: Te single-crystal ingots grown by the Czochralski method and the black allotrope of Se obtained by a vacuum distillation method at the Institute of Non-Ferrous Metals, Skawina, Poland. Both with the level of residual metallic impurities corresponding to the 4.5 - 5 N standard. The initial samples were obtained by mechanical cleaving under ambient atmospheric conditions, and their surface was covered by native oxide.
Positive and negative SIMS measurements were performed with a TOF-SIMS V instrument (*ION-TOF GmbH*, Muenster, Germany), using 30 keV Bi^+ primary ion beams. Secondary ion mass spectra were obtained using a cycle time of 120 µs to obtain the mass range up to ~ 1300 m/z. This allowed detecting the Te_x and Se_x clusters with $x \leq 10$. We used primary ion current <1.5 pA at the crater bottom 100x100 μm^2. The total fluence was less than $5x10^{12}$ Bi^+ cm^{-2} ensuring static conditions and the changes at the spectra after the second measurement in the same place were comparable with the reproducibility of the measurements.
The mode with constant primary ion current (DC mode for the TOF-SIMS V, ion current 40 pA, the crater bottom 100x100 μm^2) were used to remove surface contaminants. SurfaceLab software (6.4 version) of *ION-TOF GmbH* was used to manipulate the mass spectra. All spectra were calibrated to at least 5 peaks.

* Corresponding author: malgorzata.trzyna@gmail.com

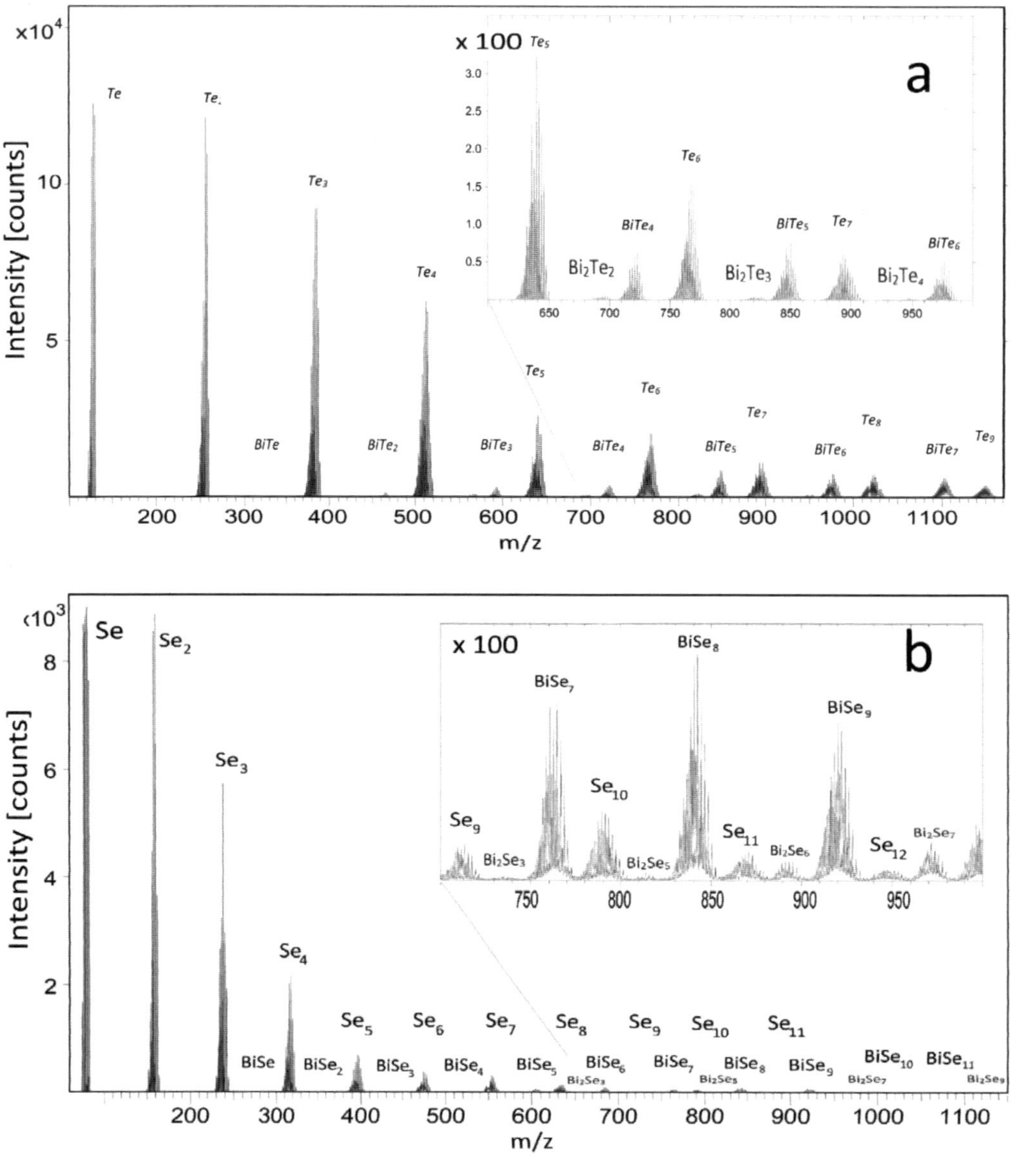

Fig.1.Negative ion mass spectra from Te and Se surfaces which were pre-cleaned with a Bi^+ at the DC mode: (a) tellurium (b) selenium.

Results

Figure 1 illustrates typical features of negative ion mass spectra from Te and Se surfaces which were pre-cleaned with a Bi^+ at the DC mode.

Both spectra contain three sequences of fragments with a typical Te or Se isotopic distribution. The sequences are shifted towards each other by 209 m/z, i.e. Bi atom mass. The first sequence is generated by $Te(Se)_x$ clusters, while the second and the third ones result from $BiTe(Se)_x$ and $Bi_2Te(Se)_x$ correspondingly. It is seen in Figure 1 that the mass spectra of secondary ions for both materials are comparable.

At the same time they are very different from the mass spectra of Se and Te clusters produced by the inert gas condensation technique [5] or laser ablation [6].

J. Becker and A. Benamar explain the completely diverse mass spectra of Se and Te by the different crystalic structures of these materials. Small cycling modules play a great role in selenium. In tellurium, these periodic structures do not exist and large clusters are easily obtained in nucleation conditions where only small selenium clusters are present [5,7].

* Corresponding author: malgorzata.trzyna@gmail.com

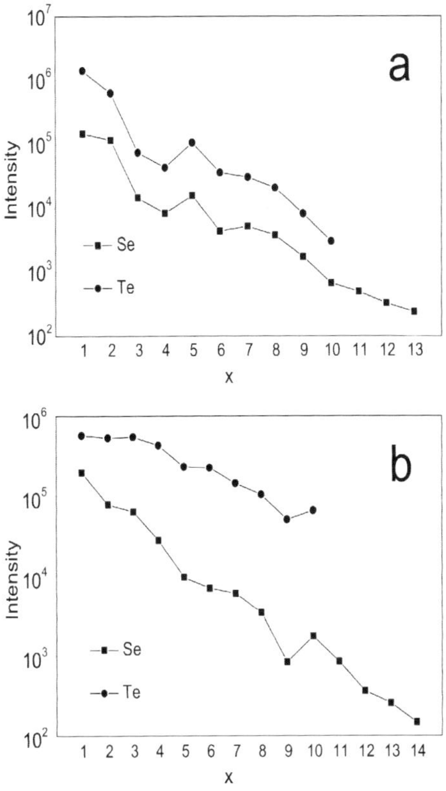

Fig. 2. The comparison of Te_x and Se_x clusters peaks intensity as function of the number x of atoms in the cluster for the sputtering by 30 keV Bi^+ primary ion beam. (a) positive ions, (b) negative ions.

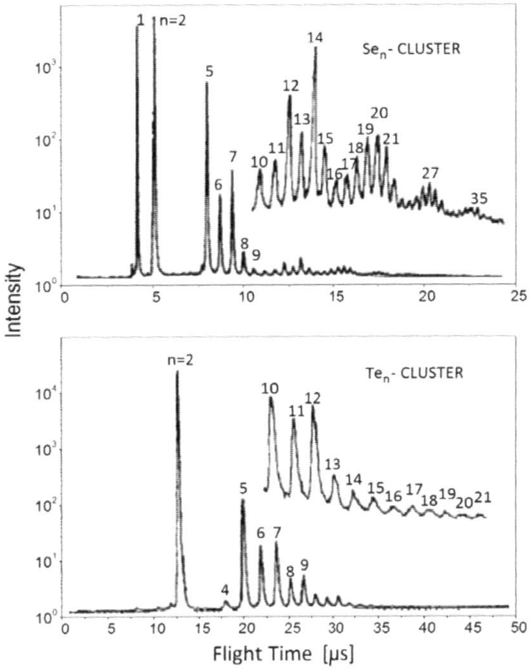

Fig.3. Typical mass-spectra of selenium- and tellurium-clusters obtained by ultraviolet photoion-photoelectron method [5].

And we must note that Se_x and Te_x clusters distribution is similar. (Fig. 2.) If we ignore curve extremes, we can see that for both positive and negative ions the general trend follows the power law in a good agreement with the results obtained for different semiconductors at sputtering by different primary ions (Ga^+ [8], Au^+, Au^-, Al^- [4], Ar^+, Cs^+, $C60^+$ [9] etc.).

This permits to suppose that in case of sputtering the mechanism of cluster formation is less dependent on the crystal structure of the starting material.

At the same time for all Se and Te clusters obtained by different methods, the maximum at x=5 is similar to our observed results. (Fig. 2.)

This maxima was also observed by J. Becker at al. They present the first mass-selected ultra- violet photoion-photoelectron coincidence spectra of Se-clusters up to Se_{25} and of Te_6 Te-clusters. Figure 3 presents these mass-spectra of Se- and Te-clusters, which was generated during the adiabatic expansion of molecular beam [5].

Similar results were also obtained by A.K. Hearley et al. The three-dimensional EDLDI method revealed the maximum for clusters formed by the laser ablation of solid samples of some elements (phosphorus, sulfur, selenium, tellurium and their mixtures) [6].

The authors using laser desorption/ionization mass spectrometry method showed the energy dependence of positive and negative ions which are formed by extension to other substances. The results prove existing the maximum of clusters, which are characteristic for the investigated materials.

B.C. Pan theoretically investigated the geometric structures and energetic of positive and neutral small tellurium clusters Te_x ($x \leq 8$), and also showed that the intensity maximum of Te_5 in the mass spectra is tied to the maximum at x=5 at the lowest dissociation energy curve of Te_x [10].

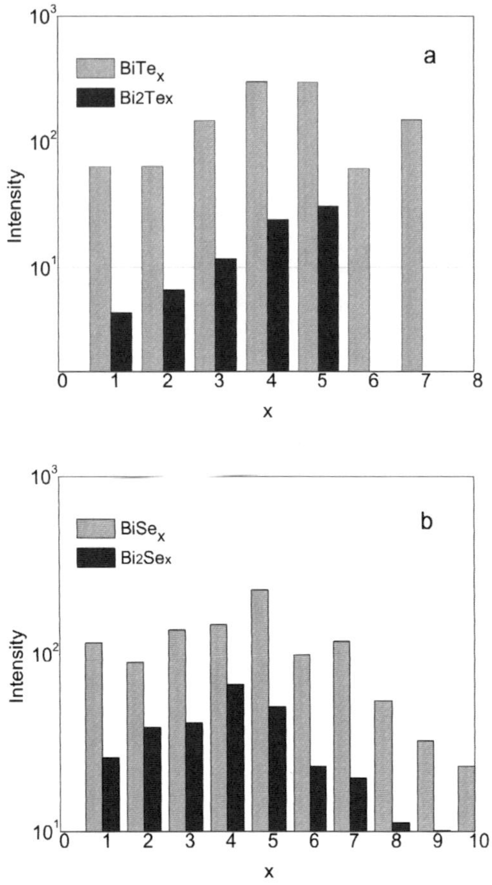

Fig.4. Comparison of peaks intensity of clusters Bi-Se and Bi-Te ions as function of the number *x* of atoms in the cluster for positive ions

Figure 4 indicates thats in addition to chalcogen clusters in mass spectra the chalcogen-bismuuth clusters can be seen. As it was demonstrated by Li and Hirokawa, the appearance of such clusters is characteristic for elements forming by atoms of a primary beam of the chemical compound or solid solution. In this case, apart from the sputtering and ionization, the alloying of surface atoms with Bi atoms takes place [8]. Though many Bi-Se and Bi-Te compounds are possible, we also reveal Bi_2Se_3 and Bi_2Te_3 as stoichiometric phase in these sequences. But on the dependences of intensity ions of Bi_2Se_x and Bi_2Te_x as function of the number x of atoms in the cluster there is no maximum for Bi_2Se_3 and Bi_2Te_3. It suggests that during the process of creating the cluster the stoichiometric complex does not appear. (Fig. 4.)

A more complete explanation of above dependences requires theoretical calculations similar to those performed for many monoatomic and polyatomic clusters of metals and semiconductors.

Conclusions

To sum up, the general view of selenium and tellurium mass-spectra is similar. Obtained clusters are comparable with the clusters obtained by different ion bombardment. However local intensity of clusters is specific for Te and Se.

The maximum corresponding to x=5 at the ion spectra Te and Se can be explained theoretically. It may indicate the greater stability of cluster structure with x = 5 in the these chalcogens.

The appearance of Bi-Se clusters is characteristic for elements forming with atoms a primary beam of the chemical compound. In this case, apart from the sputtering and ionization, the alloying of surface atoms with Bi atoms takes place.

Interactions between sputtering species and inorganic compounds involve many complicating effects. They may alter the secondary ion yield of matrix atoms and create new peaks that cause unexpected mass interferences.

Besides Bi-Se and Bi-Te clusters may find application as the building blocks for bismuth based on nanomaterials and topological insulators.

References

1. H. Zhang, C.-X. Liu, X.-L. Qi, X. Dai, Z. Fang, S.-C. Zhang, Nat. Phys. **5,** 438–442 (2009)
2. S. Lin, W. Li, Z. Chen, J. Shen, B. Ge, Y. Pei, Nat. Commun. **7,** 10287 (2016)
3. B.J.G. N. Winograd, K.E. Foley, Physic Lett. **73,** 253–255 (1979)
4. M.P. Seah, F.M. Green, I.S. Gilmore, J. Am. Soc. Mass Spectrom. **21,** 370–377 (2010)
5. J. Becker, K. Rademann, F. Hensel, Atom Mol. Clust. **235,** 233–235 (1991)
6. A.K. Hearley, B.F.G. Johnson, J.S. McIndoe, D.G. Tuck, Inorganica Chim. Acta. **334,** 105–112 (2002)
7. A. Benamar, D.I. Rayane, P. Melinon, B. Tribollet, M. Broyer, Atom Mol. Clust. **239** 237–239 (1991)
8. Z. Li, K. Hirokawa, Appl. Surf. Sci. **220,** 136–153 (2003)
9. R.E. Goacher, H. Luo, J.A. Gardella, Anal. Chem. **80,** 3261–3269 (2008)
10. B.C. Pan, Phys. Rev. B. **65,** 85407 (2002).

DOI: 10.1051/epjconf/201713302005

Shape effect in FMR of Ni-Co-Mn-In layers obtained by pulsed laser deposition

Łukasz Dubiel[1,*], *Marian* Kuzma[1], *Ireneusz* Stefaniuk[2], *Grzegorz* Wisz[3], and *Andrzej* Wal[1]

[1]Faculty of Mathematics and Natural Sciences, Department of Biophysics, University of Rzeszow, S. Pigonia 1, 35-959 Rzeszow, Poland.
[2] Faculty of Mathematics and Natural Sciences, Center for Microelectronics and Nanotechnology, University of Rzeszow, S. Pigonia 1, 35-959 Rzeszow, Poland.
[3]Faculty of Mathematics and Natural Sciences, Department of Experimental Physics, University of Rzeszow, S. Pigonia 1, 35-959 Rzeszow, Poland

Abstract. We have studied thin layers of $Ni_{50-x}Co_xMn_{50-y}In_y$ alloys on (001) Si substrate obtained by pulsed laser deposition method (PLD) using YAG Nd^{3+} laser operating at second harmonic. The target was bulk $Ni_{50-x}Co_xMn_{50-y}In_y$ (x = 5, y = 14.5) alloy prepared by induction melting of pure elements under argon atmosphere. Magnetic properties were investigated on Bruker X band EPR spectrometer (9.36 GHz) at room temperature. The magnetic resonance spectrum consists of non-symmetric lines with resonance field within wide field range (2500-4800 Gs) depending on the orientation of the static field in the plane perpendicular to the layer. Calculated spectroscopic splitting factor g = 2.09.

1 Introduction

The alloys system $Ni_{50-x}Co_xMn_{25+y}Z_{25-y}$ (Z = Sn, In, Ga, etc.), is Co-doped off-stoichiometric version of better-known Ni_2MnZ full Heusler alloys [1]. Addition of Co caused important changes in magnetism and structure [2].

The Ni_2MnZ and Ni_2CoMnZ Heusler alloys undergo martensitic phase transition (MPT) upon changing temperature and they are examples of ferromagnetic shape memory alloys [3-6]. The Ni_2MnGa undergo MPT in stoichiometric form, while Ni_2MnZ (Z = Sn, In, Ge, Sb.) exhibit MPT only in the off-stoichiometric composition [6]. Additionally, the NiMnIn based Heusler alloys in a narrow temperature range could undergo magnetically induced phase transition: from a weak-magnetic martensite to a ferromagnetic austenite [3, 7].

The advantageous properties of Heusler films have initiated study of this group of material [8, 9]. In addition the thin film technology is promising for application in nanodevices (see at [1] chapter 5). In this research, we registered ferromagnetic resonance (FMR) spectra of a NiCoMnIn layers which were prepared on Si substrate by using pulsed laser deposition method.

2 Sample preparation and experimental techniques

The NiCoMnIn films were deposited on silicon substrates by pulsed laser deposition (PLD). The off-stoichiometric bulk $Ni_{45}Co_5Mn_{35.5}In_{14.5}$ was made by arc melting [10] and was used as a target. Films were grown in PREVAC PLD system using the YAG: Nd^{3+} laser with the 532 nm (II harmonics) wavelength, 6 ns pulse time, 10 Hz repetition rate and fluence F in the range of 16 J/cm^2. The laser beam was focused on the target using a quartz lens with focal distance of 600 mm. The growth temperature T_s was 300 K. The deposition of the layers was carried out at 10^{-7} mbar vacuum.

The FMR measurements were performed on Bruker ELEXSYS E580 spectrometer using the X-band (9.36 GHz). Angular dependence was registered with the use of uniaxial goniometer at room temperature. The samples for these measurements were fixed in the microwave cavity in such a way that applied static field was perpendicular to rotating sample (Fig. 1). The *xyz* coordinate system in Fig. 1 is the system connected with a sample. The *z* is a rotation axis. The spectra were recorded as a function of angle with a 4 degrees step.

3 Results and discussion

The addition of Co and change of composition influence on the martensitic transformation and Curie temperature. [10-12]. In the $Ni_{50-x}Co_xMn_{25+y}In_{25-y}$ alloys with the 5 at.% of Co (x = 5) a T_C^A (Curie temperature in austenite phase) temperature was shifted to 343 K [10]. Therefore at the room temperature we could registered ferromagnetic resonance.

* Corresponding author: lukasz.dubiel@ifj.edu.pl

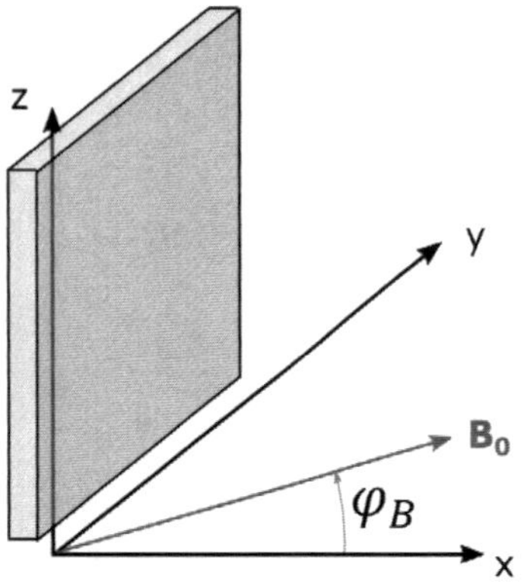

Fig. 1. Schematic orientation of film with respect to static field $\mathbf{B_0}$.

FMR spectra of the investigated sample at room temperature are shown in Fig. 2 for different angle φ_B between normal to the plane (x axis) and external field $\mathbf{B_0}$ (see Fig. 1). The spectra contain a broad and asymmetrical line of the Dyson shape [13, 14].

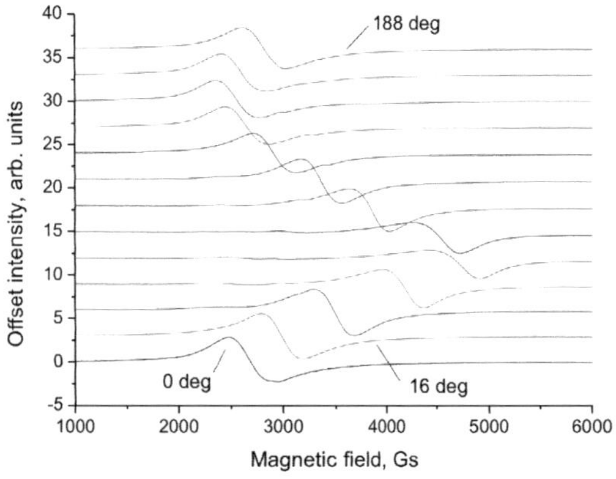

Fig. 2. The angular dependence of the FMR spectra of a film at room temperature. The spectra have been shifted vertically for better visibility.

The asymmetry FMR line suggested that investigated sample is conductive and line shape is connected with skin effect. The FMR spectra of $Ni_{50-x}Co_xMn_{50-y}In_y$ film were fitted by using derivative of Dyson function:

$$P(B) = \frac{\Delta B+\alpha(B-B_{res})}{4(B-B_{res})^2+\Delta B^2} + \frac{\Delta B+\alpha(B+B_{res})}{4(B+B_{res})^2+\Delta B^2} \quad (1)$$

where B_{res} is the resonance field, ΔB is the linewidth and α is the parameter of asymmetry which describes the absorption and dispersion ratio [14]. The examples of fitting are presented in Fig 3.

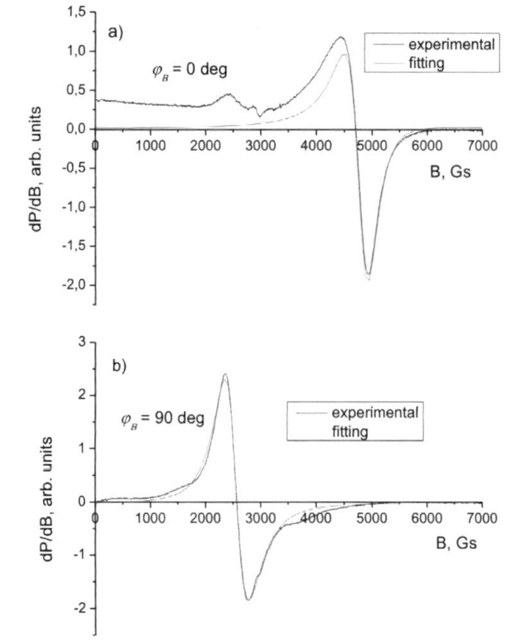

Fig. 3. The FMR signals of the NiCoMnIn film for $\varphi_B = 0$ deg and 90 deg.

It could be seen that the line shape depends on the angle between applied static field and the film (Fig. 1). The careful study of obtained spectra reveals overlapping signals. The analysis of FMR parameters was done in this paper for the strongest asymmetric line, whereas the complexity of spectrum will be investigated elsewhere. The parameters of fitted lines are collected in table 1.

Table 1. The fitting parameter of the EPR signals for $\varphi_B = 0$ and 90 deg.

φ_B [deg]	B_{res} [Gs]	ΔB [Gs]	asymmetry parameter
0	2534	697	-1,39
90	4803	731	0,38

The ferromagnetic resonance field is described by Landau-Lifshitz equation (LL equation)

$$\dot{\mathbf{M}} = -\gamma(\mathbf{M}\times\mathbf{B}) - \alpha\frac{\gamma}{\mathbf{M}}[\mathbf{M}\times(\mathbf{M}\times\mathbf{B})] \quad (2)$$

where $\mathbf{M}$ is magnetization vector, $\mathbf{B}$ is the effective magnetic field, γ is the gyromagnetic ratio and α is Gilbert damping parameter associated with a relaxation mechanisms of vector $\mathbf{M}$.

In the case of a ferromagnetic resonance of a flat plane with external magnetic field B_0 perpendicular or parallel to the plane the solution of LL equation reduced to two Kittel equation [16]:

DOI: 10.1051/epjconf/201713302005

$$\frac{\omega}{\gamma} = [B_{\parallel}(B_{\parallel} + \mu_0 M)]^{1/2} \quad (3)$$
$$\frac{\omega}{\gamma} = B_{\perp} + \mu_0 M \quad (4)$$

where $B_0 = B_{\parallel}$ is a resonance field in plane case and $B_0 = B_{\perp}$ corresponds to of plane case. Taking into account that $\gamma = \frac{g\mu_B}{\hbar}$, these two equations could be written as:

$$\hbar\omega = g\mu_B[B_{\parallel}(B_{\parallel} + \mu_0 M)]^{1/2} \quad (5)$$
$$\hbar\omega = g\mu_B(B_{\perp} + \mu_0 M) \quad (6)$$

Comparing right sites of equations (3) and (4) we obtain the equation

$$x^2 - (B_{\parallel} + 2B_{\perp})x + B_{\perp}^2 - B_{\parallel}^2 = 0 \quad (7)$$

where $x \equiv \mu_0 M$.

Inserting to (7) experimental values $B_{\parallel} = 2534$ Gs and $B_{\perp} = 4803$ Gs given by fitting curves (see Tab. 1), the equation is reduced to

$$x^2 - (2534 + 2 \cdot 4803)x + 4803^2 - 2534^2 = 0 (7)$$

From the set of eq. (5) and (6) the g-factor and effective saturation magnetization $\mu_0 M$ could be calculated. The solution are: $g = 2.09$ and $\mu_0 M = 1576$ Gs. The value of g-factor is similar to g-factor $g = 2.01$ listed by Kittel (see Table I [17]) for Heusler alloys.

3 Conclusion

The layers of $Ni_{50-x}Co_xMn_{50-y}In_y$ have been grown on (100) Si substrate by PLD method. Angular dependence of FMR spectra was registered at room temperature. The obtained FMR spectra were fitted by Dyson function. The experimental FMR lines exhibited a strong angular dependence and shape anisotropy with resonance field changed from 2600 Gs (for parallel geometry) until 4800 Gs (for perpendicular geometry). Calculation of g-factor (g = 2.09) provides a typical value for Heusler alloys [17].

References

1. C. Felser, A. Hirohata, *Heusler Alloys - Properties, Growth, Applications* (Switzerland, 2016)
2. S. Fabbrici, G. Porcari, F. Cugini, M. Solzi, J. Kamarad, Z. Arnold, et al., Entropy **16**, 2204 (2014)
3. T. Krenke, E. Duman, M. Acet, E.F. Wassermann, X. Moya, L. Mañosa, et al., Nat. Mater. **4,** 450 (2005)
4. T. Krenke, M. Acet, E.F. Wassermann, X. Moya, L. Mañosa, A. Planes, Phys. Rev. **B. 73**, 174413 (2006)
5. Y. Sutou, Y. Imano, N. Koeda, T. Omori, R. Kainuma, K. Ishida, et al., Appl. Phys. Lett. **85**, 4358 (2004)
6. C.M. Li, H. Bin Luo, Q.M. Hu, R. Yang, B. Johansson, L. Vitos, Phys. Rev. **B 86**, 214205 (2012)
7. P. Klaer, H.C. Herper, P. Entel, R. Niemann, L. Schultz, S. Fähler, et al., Phys. Rev. **B 88**, 174414 (2013)
8. S. Rios, I. Karaman, X. Zhang, Appl. Phys. Lett. **96**, 173102 (2010)
9. R. Niemann, O. Heczko, L. Schultz, S. Fähler, Appl. Phys. Lett. **97**, 222507 (2010)
10. W. Maziarz, Solid State Phenom. **186**, 251 (2012)
11. W. Ito, Y. Imano, R. Kainuma, Y. Sutou, K. Oikawa, K. Ishida, Metall. Mater. Trans. A Phys. Metall. Mater. Sci. **38**, 759 (2007)
12. W. Maziarz, P. Czaja, M.J. Szczerba, J. Przewoznik, C. Kapusta, A. Zywczak, et al., J. Magn. Mater. **348**, 8 (2013)
13. G. Feher, A.F. Kip, Electron spin resonance absorption in metals. I. Experimental, Phys. Rev. **98**, 337 (1955)
14. J.P. Joshi, S. V. Bhat, On the analysis of broad dysonian electron paramagnetic resonance spectra, J. Magn. Reson. **168**, 248 (2004)
15. J. Dubowik, K. Załęski, H. Głowiński, I. Gociańska, Phys. Rev. B. **84**, 184438 (2011)
16. C. Kittel, Phys. Rev. **73**, 155 (1948)
17. C. Kittel, Phys. Rev. **76**, 480 (1949)

Structure and composition of layers of Ni-Co-Mn-In Heusler alloys obtained by pulsed laser deposition

Grzegorz Wisz[1,*], *Piotr* Sagan[2], *Ireneusz* Stefaniuk[3], *Bogumil* Cieniek[1], *Wojciech* Maziarz[4], and *Marian* Kuzma[2]

[1]Faculty of Mathematics and Natural Sciences, Department of Experimental Physics, University of Rzeszow, Pigonia 1, 35-959 Rzeszow, Poland
[2]Faculty of Mathematics and Natural Sciences, Department of Biophysics, University of Rzeszow, Pigonia 1, 35-959 Rzeszow, Poland
[3]Faculty of Mathematics and Natural Sciences, Center for Microelectronics and Nanotechnology, University of Rzeszow, Pigonia 1, 35-959 Rzeszow, Poland
[4]Institute of Metallurgy and Materials Science of the Polish Academy of Science, 25 Reymonta Str., 30-059 Krakow, Poland

Abstract. In present work we were analysing thin layers of Ni-Co-Mn-In alloys, grown by pulsed laser deposition method (PLD) on Si, NaCl and glass substrates. For target ablation the second harmonics of YAG:Nd^{3+} laser was used. The target had the composition $Ni_{45}Co_5Mn_{34.5}In_{14.5}$. The morphology of the layers and composition were studied by electron microscopy TESCAN Vega3 equipped with microanalyzer EDS – Easy EdX system working with Esprit Bruker software. The X-ray diffraction measurements (XRD), performed on spectrometer Bruker XRD D8 Advance system, reveals Ni_2-Mn-In cubic phase having lattice constant a = 6.02Å.

1 Introduction

A phenomenon of a giant magnetoresistance (GMR) and a phenomenon of a tunnel magnetoresistance (TMR) in structures of a magnetic layer/layer of metal (or a dielectric) /magnetic layer are used for spintronics purposes. Heusler alloys which base on Mn-Ge [1, 2], Mn-Si [3] or CoMnSi [4] are very promising. Nevertheless, other multifunctional Heusler alloys having nonordinary properties like high perpendicular magnetic anisotropy, low damping constant or half-metallicy could be useful in spintronic as well [5]. The growth and investigation of thin layers of such materials are highly desired step for future applications in field of spintronics. NiMnIn belongs to the separate group of Heusler alloys called metamagnetic alloys (NiMnGa, NiMnSn, NiMnIn) [6]. The strong magnetic and structural nonhomogeneity in this compound may be the source of GMR[7].

In this paper we grown thin layers from bulk $Ni_{50-y}Co_yMn_{50-x}In_x$ (y=5,

* Corresponding author: gwisz@ur.edu.pl

DOI: 10.1051/epjconf/201713302006

x=14.5). Addition of Co into NiMnIn moves the Curie temperature to the region of room temperature or higher [8]. We have applied pulsed laser deposition method (PLD) [9] for layers fabrication, which is a low temperature method what ensures the abrupt interfaces [10]. Moreover, PLD is non equilibrium method what can maintain the composition of the target in the layer [11].

2 Experimental

The NiCoMnIn films on silicon, glass and NaCl substrates were deposited by PLD using the YAG: Nd^{3+} laser with the 532 nm (II harmonics) wavelength, 6 ns pulse time, 10Hz repetition rate and fluence F in the range 16 J/cm^2. The bulk $Ni_{50-x}Co_xMn_{35.5}In_{14.5}$ was used as a target [12]. The laser beam was focused on the target using a quartz lens with focal distance of 600 mm. The growth temperature T_s was kept at 20°C – 200°C and the deposition of the layers was carried out at 10^{-7} mbar vacuum. The samples list and technological parameters contains Table 1.

Table 1. List of PLD samples and technological parameters.

No.	Substrate	Substrate temperature [K]	Process time [min]	Vacuum [mbar]	Analysis
P1	Si, Glass, Al_2O_3	293	15	$1.9*10^{-7}$	
P2	Si, NaCl, Glass	305	20	$2.4*10^{-7}$	EDS, SEM
P3	Si, NaCl, Glass	473	20	$1.6*10^{-7}$	
P4	Si, NaCl, Glass	302	10	$1.48*10^{-7}$	
P5	Si, NaCl, Glass	303.6	30	$1.48*10^{-7}$	
P6	Si, NaCl, Glass	301.2	30	$1.01*10^{-7}$	EDS, SEM, XRD
P7	Si, NaCl, Glass	301.1	10	$7.07*10^{-8}$	EPR
P8	Si, NaCl, Glass, N2"Wh	373	30	$1.16*10^{-7}$	XRD

Composition and morphology of the films were investigated by scanning electron microscope Tescan Vega3, which was equipped in energy dispersive X-ray spectroscopy (EDS) analyser. The X-ray diffraction study was performed on spectrometer Brucker XRD D8 Advance System.

3 Results and discussion

3.1 Morphology and composition

The obtained layers are very smooth by visual inspection. SEM microscopy observations (Fig. 1b) confirm this statement. The thickness of the sample was determined on about 75 nm based on the edge of the breaking sample shown in Fig. 1a. Many regular drops were observed on the surface. The size of drops is 0.5 - 2 µm. Such existence of drops is typical for layers obtained by PLD. In Fig. 1c a very regular shape (as ball) of one exceptional great drop is presented.

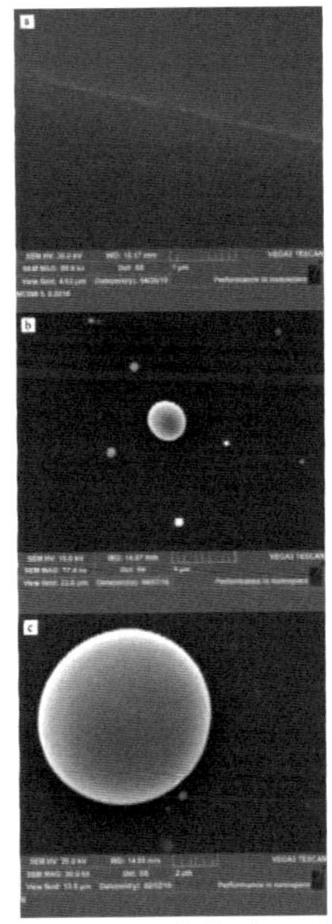

Fig. 1. SEM images of the layer P6 at small (a) and great magnification (b). Shape of the great drop (c).

Composition of layers was measured by EDS at low accelerating voltage due to small thickness of the films. In Fig. 3. results obtained at 5 kV and 10kV were compared. One can notice fair convergence of both results. In Table 2 the results of particular measurements were collected. From these data the average value was calculated. The thickness of layer is much smaller than the excitation area therefore the amount of Si is greater and accuracy of measure is lower. The composition of drop (shown in Fig. 1c) is determined with better accuracy while the diameter is greater than the size of excitation area (small amount of Si substrate in account).

Table 2. Composition at at. % of the layer P6 and the drop in it.

	5kV		10kV		Average composition of the layer
	Layer P6		Layer P6	Drop	
	without Si	with Si	with Si	with Si	
Si	-	37.26	84.93	1.52	61.10
Ni	69.57	67.41	65.83	54.73	67.60
Co	1.71	1.80	2.85	1.70	2,12
Mn	22.62	23.37	24.95	34.33	23.65
In	6.09	7.43	6.44	9.23	6.65

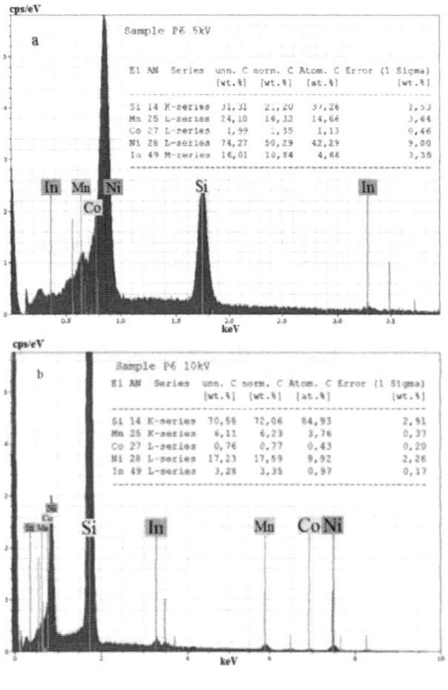

Fig. 2. Composition measured at 5 kV (a) and at 10 kV (b).

3.2 XRD analysis

In Fig. 3. X-ray diffraction patterns for samples P6 and P8 deposited on (001)Si at substrate temperature 301,2K and 373K respectively were shown. The diffraction lines are very weak due to thinness of layers. Only the reflex from Si substrate (at $2\theta = 69.325^{o}$) is very strong therefore, the intensity axis is given with logarithmic scale. Both spectra are characteristic for polycrystalline samples and do not differ considerably, so we analyse one of them: for sample P6 which was grown at room temperature. The 2θ values of diffraction lines for this sample (P6) are presented in Fig. 4. and in Table 3.

In Table 3 the parameters (2θ, intensity and hkl) of diffraction lines are collected for experimental spectrum from Fig. 4. and for more possible phases of metallic Ni (cubic), Mn (cubic) and In (cubic). Moreover there are diffraction lines of stoichiometric Ni_2MnIn and nonstoichiometric $Ni_2Mn_{1-x}In_x$ x=0.05. The best relation of some experimental diffraction lines to the database lines is for stoichiometric Ni_2MnIn. A small amount of one component cubic Ni and Mn should be considered as well.

The system of the diffractometer identify the spectrum as Ni_2MnIn using ICDD PDF-2 2012 database (see lower part in Fig. 4. and Fig. 5.c). This compound relates to the phase $Ni_{50}Mn_{50-x}In_x$, x=25 analysed by T. Krenke et all. [13] (Fig. 5b). Lattice constant for our phase is a = b = c = 6.02Å whereas, Krenke determined the lattice constant for $Ni_{50}Mn_{50-x}In_x$, x=25 a = b = c = 6.071Å. This discrepancy can be explained by nonstoichiometry. By replacing In with Mn the lattice constant of the austenitic phase become smaller. Taking into account the dependence of lattice constant on composition (see Fig. 6.) the composition $Ni_2Mn_{1+y}In_{1-y}$ for lattice constant 6.02Å is $y = 0.22 \pm 0.02$ and depends on magnetic ordering [14] what is considered in accuracy Δy. This value relates to x = 19.5.

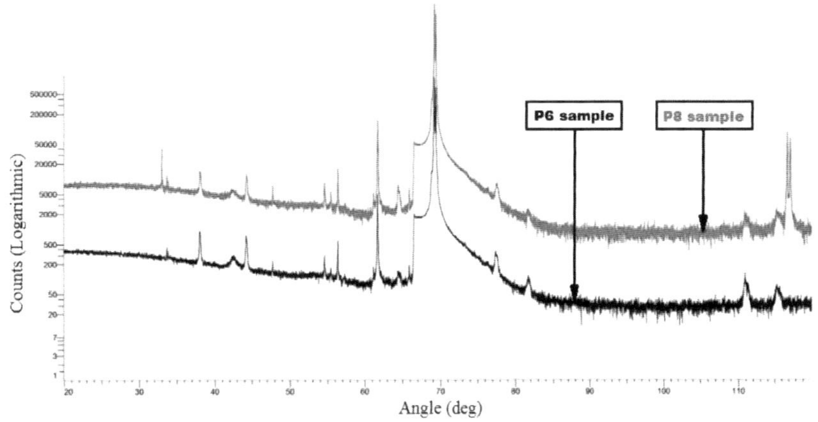

Fig. 3. X-ray diffraction patterns for samples P6 and P8.

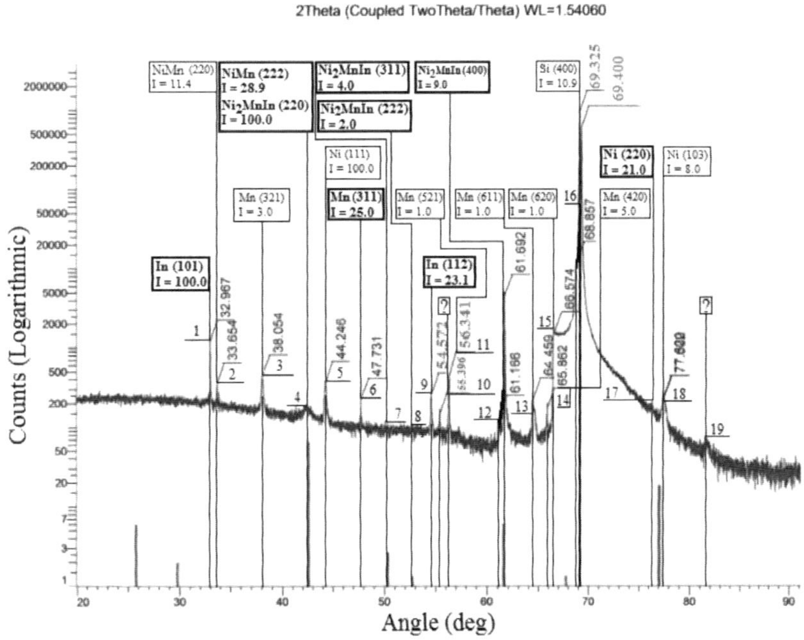

Fig. 4. X-ray diffraction pattern for sample P6 and the scheme of reference diffraction lines of phases Ni(cub), Mn(cub), In, NiMn, Ni_2MnIn (according to the Table 3).

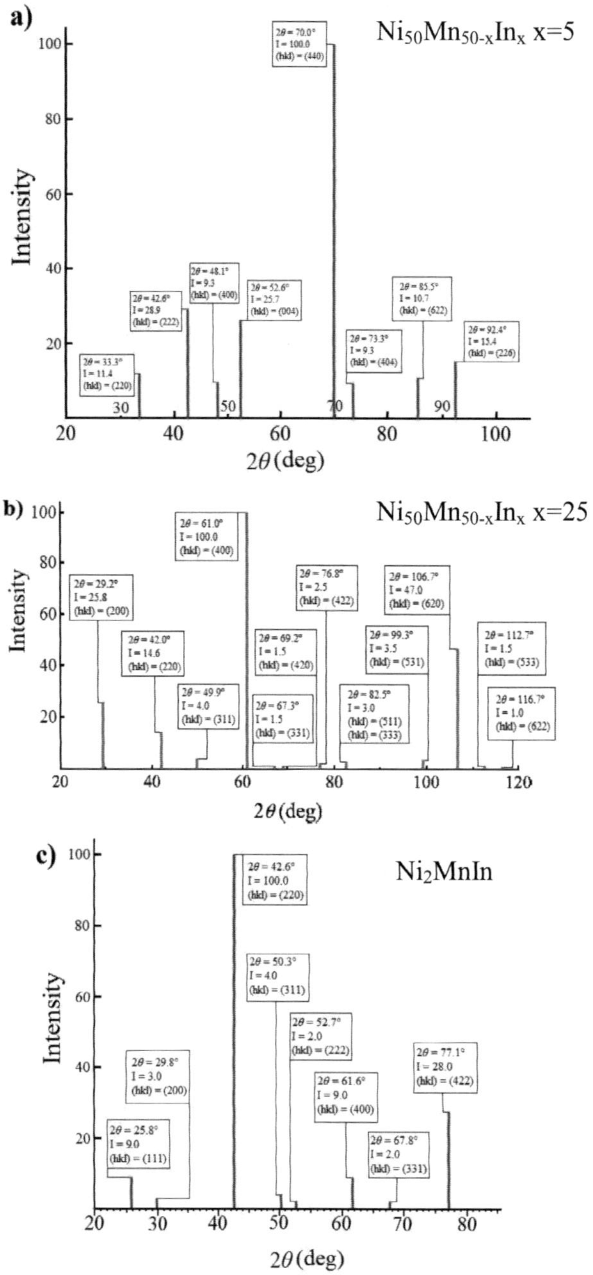

Fig. 5. X-ray diffraction lines of $L1_0$ phase of $Ni_{50}Mn_{50-x}In_x$, x=5 (according to Fig. 6. of [13]) (a), $L2_1$ phase of $Ni_{50}Mn_{50-x}In_x$, x=25 (according to Fig. 3. of [13]) (b), $L2_1$ phase of Ni_2MnIn, (according to ICDD PDF-2 database) (c).

 DOI: 10.1051/epjconf/201713302006

Table 3. Relations of the position of the database XRD lines of metallic Ni,Mn,In and Ni_2MnIn, $Ni_2Mn_{1-x}In_x$ x=0.05 compounds with experimental diffraction pattern of the sample P6.

Experimental			ASTM						database.iem.ac.ru/mincryst/			ICDD PDF-2 database, pattern: PDF 00-040-1208			Krenke et all. [13]		
			CAS num: 7440-02-0			CAS num: 7439-96-5											
			Ni(c)			Mn(c)			In(c)			Ni2MnIn			$NiMnIn_x$ x=0.05		
line	2Θ	I	2Θ	I	hkl	2Θ	I	hkl	2Θ	I	hkl	2Θ	I	hkl	2Θ	I	hkl
1.	32.967	m							33.010	100	(101)						
2.	33.654	w	37.735	3.0	(321)										33.3	11.4	(220)
3.	38.054	w															
4.	42.500	vw										42.570	100	(220)	42.6	28.9	(222)
5.	44.246	w	44.505	100	(111)												
6.	47.731	w				47.728	25.0	(311)									
7.	50.300	vw										50.315	4.0	(311)			
8.	52.700	vw										52.716	2.0	(222)	52.6	25.7	(004)
9.	54.572	w							54.563	23.1	(112)						
10.	55.396	vw	Weak, unidentified line.														
11.	56.341	m				56.502	1.0	(521)									
12.	61.166 61.692	s										61.617	9.0	(400)			
13.	64.459	w				64.378	1.0	(611)									
14.	65.862	w				66.149	5.0	(420)									
15.	66.574	m				66.292	1.0	(620)									
16.	68.857 69.325 69.400	vs							69.250	11.0	(202)						
17.	76.400	vw	76.366	21.0	(220)												
18.	77.600	w	78.000	8.0	(103)												
19.	81.600	vw	Weak, unidentified line.														

vw - very week w - week m - medium s - strong vs - very strong

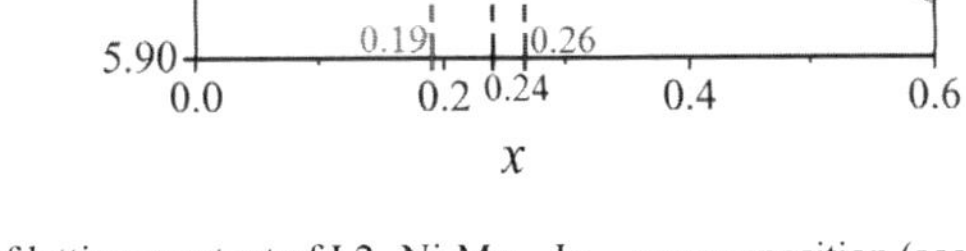

Fig. 6. Dependence of lattice constant of $L2_1$ $Ni_2Mn_{1+y}In_{1-y}$ on composition (according to Fig.2. of [14]) for different magnetic phases: ferromagnetic (FM), antiferromagnetic (AFM) and disordered (DIS).

4 Conclusions

1. Thin layers (~75 nm) on Si substrate have been grown by PLD method from target $Ni_{45}Co_5Mn_{35,5}In_{14,5}$.
2. EDS study revealed that the films have composition similar to the composition of target. However, the excess of Ni in respect to Mn were observed.
3. XRD measurments point on $Ni_2CoMnIn$ cubic phase with lattice constant a = 6.02 Å. This correspond to the composition x = 0.22 of the $L2_1$ $Ni_2Mn_{1+x}In_{1-x}$.

References

1. V.A. Oksenenko, L.N. Trofimova, Yu.N. Petrov, Y.V. Kudryavtsev, J. Dubowik, Y.P. Lee, Journal of Applied Physics **99**, 063902 (2009)
2. A. Sugihara, K. Suzuki, S. Mizukami, T. Miyazaki, J. Phys. D: Appl. Phys. **48**, 164009 (2015)
3. S. Zhou, K. Potzger, G. Zhang, A. Mücklich, F. Eichhorn, N. Schell, R. Grötzschel, B. Schmidt, W. Skorupa, M. Helm, J. Fassbender, Phys. Rev. B **75** 085203 (2007)
4. R. Fetzer et al, J. Phys. D: Appl. Phys. **48**, 164002 (2015)
5. S. Mizukami, A. Serga, J. Phys. D: Appl. Phys. **48**, 160301 (2015)
6. R. Kainuma, K. Oikawa, W. Ito, Y. Sutou, T. Kanomata, K. Ishida, J. Mater. Chem. **18**, 1873-1842 (2008)
7. I. Dubenko, A.K. Pathak, S. Stadler, N. Ali, Phys. Rev. B **80**, 092408 (2009)
8. H.C. Xuan, F.H. Chen, P.D. Han, D.H. Wang, Y.W. Du, Intermetallics **47**, 31-35 (2014)
9. D. B. Chrisey, G.K. Hubler, *Pulsed Laser Deposition of Thin Films,* Wiley & Sons, New York, 1994
10. S. Rai, M.K. Tiwari, G.S. Lodha, M.H. Hodi, M.K. Chattopadhyay, S. Majumdar, S. Gardelis, Z. Viskadourakis, J. Giapintzakis, R.V. Nandedkar, S.B. Roy, P. Chaddah, Phys. Review B **73**, 0354017 (2006)
11. J. T. Cheung and H. Sankur, CRC Crit. Rev. Solid State Matter. Sci. **15**, 63 (1988)
12. W. Maziarz, Solid State Phenomena **186**, 251-254 (2012)
13. T. Krenke, M. Acet, E.F. Wassermann, Phys. Rev. B **73**, 174413 (2006)
14. Ch.M. Li, H.B. Luo, Q.M. Hu, R. Yang, B. Johansson, L. Vitos, Phys. Rev. B **86**, 214205 (2012)

DOI: 10.1051/epjconf/201713303001

Two-photon absorption of $Tl_{1-x}In_{1-x}Sn_xSe_2$ nanocrystallites

Galyna Myronchuk[1], *Oleg* Parasyuk[2], *Ludmila* Piskach[2], *Nasser Saleh* Alzayed[3], *Serhii* Prokhorenko[4], *Michal* Piasecki[5], *Iwan* Kityk[6]

[1] Department of Solid State Physics, Eastern European University, 13 Voli Ave., Lutsk 43025, Ukraine
[2] Department of Inorganic and Physical Chemistry, Eastern European National University, 13 Voli Ave., Lutsk 43025, Ukraine
[3] Physics and Astronomy Department, College of Science, King Saud University, P.O. Box 2455, Riyadh 11451, Saudi Arabia
[4] Center for Microelectronics and Nanotechnology. University of Rzeszow, 35-959 Rzeszów, Poland
[5] Institute of Physics, J. Dlugosz University, Armi Krajowej 13/15, Czestochowa, Poland
[6] Electrical Engineering Department, Czestochowa University of Technology, Armi Krajowej 17, Czestochowa, Poland

Abstract. Novel materials for the infrared two-photon absorption — $Tl_{1-x}In_{1-x}Sn_xSe_2$ single crystals (x=0.1,0.2) were grown. Two-photon absorption (TPA) was studied at CO_2 laser wave-length 9.4μm with pulse duration 1μs. The studies were performed at different temperatures and for the nanocrystallite sizes varying within the 7–200 nm. The studies have shown that the TPA may be enhanced during the decrease of the nanocrystallite sizes below 50–60 nm. There exists also some critical x value at which the TPA value begin substantially to increase. The studied nanocrystallites are relatively stable to the infrared laser treatment and are not hygroscopic which allow to use them in different IR optoelectronic devices.

1 Introduction

The search of materials for infrared (IR) light detection in the mid infrared spectral range, (i.e. about 10μm) is very actual task [1–5]. Almost all the studies are related with the use of quantum dot IR photodetectors QDIPs [6]. All these researches require a necessity to use sophisticated and expansive quantum dot-in-a-well technology. However, there exists still a large amount of materials in which due to the processes of additional contribution of the IR induced phonons, including anharmonic ones [7], there occur large reserves for enhanced nonlinear optical susceptibilities. The latter may give substantial contribution to the nonlinear optical effects of the third order [8, 9] which present an active area of research in recent years.

In this paper we present a study study the two-photon absorption (TPA) in the IR wavelength spectral range for the $Tl_{1-x}In_{1-x}Sn_xSe_2$ nanocrystallites (x=0.1, 0.2) using as a fundamental CO_2 laser with wavelength 9.4μm and pulse duration 1μs. The studies were performed at different temperatures and for the nanocrystallite sizes varying within the 7–200 nm

2 Experimental

The titled single crystals were fabricated by Bridgman–Stockbarger method using solid solution range varying within the 0–20 mol.% $SnSe_2$. The technological process was proceed in a shaft-type similarly to the method described in Ref.[10]. Finally the such-grown grown single crystals (with diameter 9 mm and length about 30 mm) were black and possessed a good layered structure which allow to perform their cleaving.

The specimens for linear spectra and two-photon absorption of the $Tl_{1-x}In_{1-x}Sn_xSe_2$ crystals were cleaved from the relatively perfect middle part of the single-crystalline ingot and were shaped as thin plates of thickness varying within the 0.08–0.1 mm. For the nanocrystallites preparation we have used a method of crystal milling with additional acoustical milling and size separation as described in Ref.[11]

3 Results and discussion

The NC sizes and their size distribution was controlled by JEOL TEM microcopy. Typical NC size distributions is presented in Fig. 1.

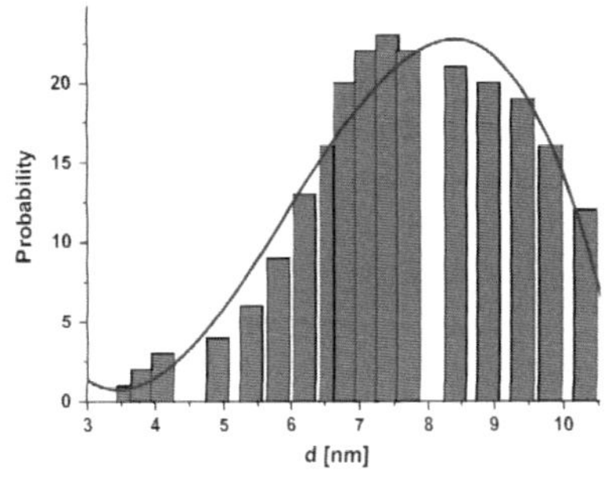

Fig. 1. Typical size distribution of the nanocrystals following the TEM analysis for the series with the lowest sizes

* Corresponding author: siergiej.prokhorenko@gmail.com

One can see a huge size dispersion which in addition shows substantial asymmetry. However, the such distribution allows to perform analysis of their optical and nonlinear optical properties.

In Fig. 2 are presented the dependences of the TPA versus the NC sizes at two temperatures: at room (ambient) temperature and the LNT. One can see a drastic increase of the TPA at average sizes below 100 nm at LNT and at below 50 nm at RT. Such differences confirms substantial role of the phonon subsystem for the such kind of the NC. At the same time the enhancement of the TPA was up to four times, which allow to predict that the nano-confined effects favoring high local electric field are dominated here.

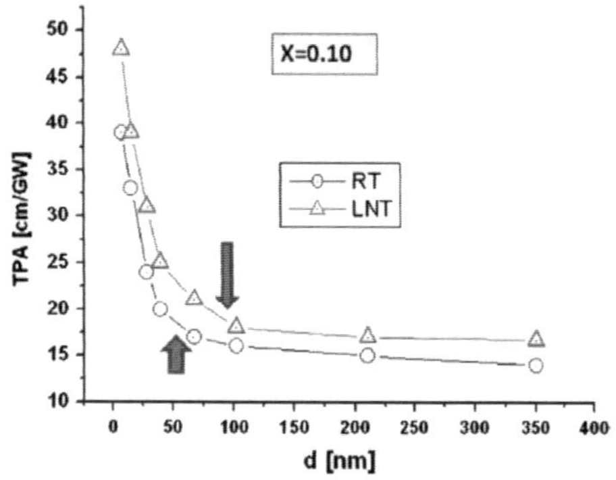

Fig. 2. Size dependence of the TPA for the samples with content x= 0.10 at RT and LNT temperatures

To explore an influence the nanocrystal composition on the TPA behavior in Fig. 3 are presented the features for x=0.20. Here the general TPA dependences versus the sizes is almost the same. However, in this case the role of the temperature is not so curial. So one can predict that in this case the local disorder defining the local trapping levels is prevailing. It is necessary to add that the illumination of the studied compounds by the corresponding laser pulses did not change the TPA more than 0.2%.

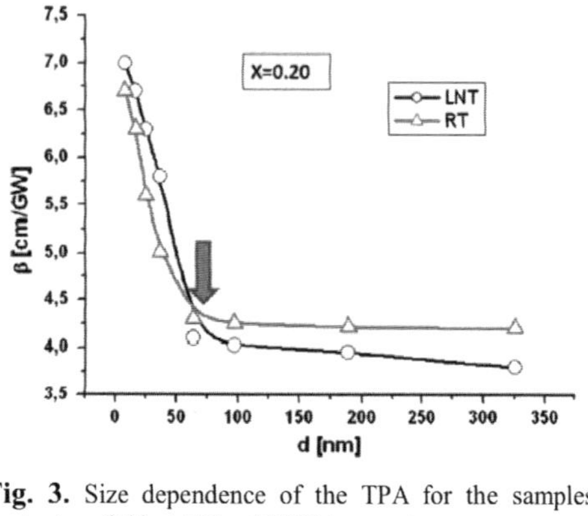

Fig. 3. Size dependence of the TPA for the samples with content x= 0.10 at RT and LNT temperatures.

The studies of the two-photon absorption for the $Tl_{1-x}In_{1-x}Sn_xSe_2$ single crystals (x= 0.1, 0.2) and corresponding nanocrystals at 9.4μm has shown high value of the TPA coefficients for the diagonal tensor components which achieve the value up to 50 cm/GW. The maximal TPA value achieved for diagonal tensor components occurring at the x= 0.1. It is principal that the TPA decreases with decreasing temperature. Generally it shows sub-stantially non-monotonous dependence versus the Tl content. This fact is a result of two different contributions: the long-range ordering and the mentioned trapping levels. The essential crucial temperature sensitivity behavior may be caused by additional phonon contribution to these localized trapping levels. It is crucial that the effect was stable for the laser beams varying within the wide range of beam spot diameters: from 5 mm up to 3 cm.

The achieved parameters of nonlinear optical effects derived for the wavelengths corresponding to the CO_2 laser illumination allows to predict wide applications of $Tl_{1-x}In_{1-x}Sn_xSe_2$ single crystals as materials for nonlinear optical transformation of the infrared laser beams which is crucial in the infrared LIDAR (Light Detection and Ranging) systems. In particular, the crystals may be effectively used in the photonic gratings [12]. It is necessary to add that during multi-time IR induced treatment the observed photoinduced changes were completely reversible which confirm their high reproduction potential [13–16].

Conclusions

During exploration of the TPA versus the sizes was discovered a began of drastic increase the TPA at nanocrystal average sizes about 100 nm at LNT and at about 50 nm at RT. Such differences confirms substantial role of the phonon subsystem for the such kind of the NC. At the same time the enhancement of the TPA was up to four times, which allow to predict that the nano-confined effects favoring high local electric field are dominated here.

References

1. Z.H. Chen, O. Baklenov, E.T. Kim, I. Mukhametzhanov, J. Tie, A. Madhukar, Z. Ye, J.C. Campbell, J. Appl. Phys. **89,** 4558 (2001)
2. P. Karasinski, Opt. Appl. **34**, 467 (2004)
3. P. Aivaliotis, L.R. Wilson, E.A. Zibik, J.W. Cockburn, M.J. Steer, H.Y. Liu, Appl. Phys. Lett. **91**, 013503 (2007)
4. H. Lim, S. Tsao, W. Zhang, M. Razeghi, Appl. Phys. Lett. **90,** 131112 (2007)
5. H.C. Liu, B. Aslan, M. Korkusinski, S.J. Cheng, P. Hawrylak, Infrared Phys. Technol. **44**, 503 (2003)
6. P. Aivaliotis, E.A. Zibik, L.R. Wilson, J.W. Cockburn, M. Hopkinson, N.Q. Vinh, Appl. Phys. Lett. **92**, 023501 (2008)

7. Zhiting Tian, Jivtesh Garg, Keivan Esfarjani, Takuma Shiga, Junichiro Shiomi, Gang Chen, Phys. Rev. B **85**, 184303 (2012)
8. Victor Romero-Rochin, Richard M. Koehl, Ciaran J. Brennan, Keith A. Nelson, J. Chem. Phys. **111**, 3559 (1999)
9. O.U. Fa, H.E. Minggaoo, W.U. Fugen, J. Nonlinear Opt. Phys. Mater. **10**, 389 (2001)
10. M.Yu. Mozolyuk, L.V. Piskach, A.O. Fedorchuk, I.V. Kityk, I.D. Olekseyuk, O.V. Parasyuk, J. Alloys Compd. **509**, 2693 (2011)
11. A. Majchrowski, I.V. Kityk, T. Lukasiewicz, A. Mefleh, S. Benet, Opt. Mater. **15 (1)**, 51–58 (2000).
12. E. Gondek, P. Karasinski, Opt Laser Techno **l48**, 438 (2013)
13. K.J. Plucinski, M. Makowska-Janusik, A. Mefleh, I.V. Kityk, V.G. Yushanin, Mater Sci EngB **64**, 88 (1999)
14. I.V. Kityk, M. Demianiuk, A. Majchrowski, J. Ebothe, P. Siem-ion, J. Phys.: Condens. Matter **16**, 3533 (2004)
15. G.E. Davydyuk, M. Piasecki, O.V. Parasyuk, G.L. Myronchuk, A.O. Fedorchuk, S.P. Danylchuk, L.V. Piskach, M.Yu. Mozolyuk, N. AlZayed, I.V. Kityk Optical Materials **35**, 2514 (2013)
16. G. L. Myronchuk, G. E. Davydyuk, O. V. Parasyuk, O. Y. Khyzhun, R. A. Andrievski, A. O. Fedorchuk, S. P. Danylchuk, L. V. Piskach, M. Y. Mozolyuk J Mater Sci: Mater Electron **24**, 3555 (2013)

Ion implantation of CdTe single crystals

Tomasz Wiecek [1], *Volodymir* Popovich[2], *Mariusz* Bester[3], *Marian* Kuzma[3]

[1]Physics Chair, Rzeszow University of Technology, al. Powstańców Warszawy 12, 35-959 Rzeszow, Poland
[2] Department of Fundamental Technologies, Ivan Franko Drogobych State Pedagogical University, Ivan Franko 24, 82100 Drogobych, Ukraine
[3]Biophysics Chair, Rzeszow University, S. Pigonia 1, 35-959 Rzeszow, Poland

Abstract. Ion implantation is a technique which is widely used in industry for unique modification of metal surface for medical applications. In semiconductor silicon technology ion implantation is also widely used for thin layer electronic or optoelectronic devices production. For other semiconductor materials this technique is still at an early stage. In this paper based on literature data we present the main features of the implantation of CdTe single crystals as well as some of the major problems which are likely to occur when dealing with them. The most unexpected feature is the high resistance of these crystals against the amorphization caused by ion implantation even at high doses (10^{17} $1/cm^2$). The second property is the disposal of defects much deeper in the sample then it follows from the modeling calculations. The outline of principles of the ion implantation is included in the paper. The data based on RBS measurements and modeling results obtained by using SRIM software were taken into account.

1 Introduction

Doping of single crystals of CdTe is a troublesome task in AIIBVI devices technology due to the segregation effects , the presence of compensating defects in the crystal and due to disturbing of stoichiometry [1].There are three main methods of doping of CdTe crystals : the first being the addition of species during growth, the second one reckoning on the diffusion of dopantss using annealing, and the third one which is the ion implantation. The two first methods are extensively experimentally investigated for wide amount of dopant elements (see Chapter B5 in [1]). Contrarily, the ion implantation has been studied for two dopants only : As and Bi . The aim of such studies was type conversion for p-n junction formation. Ion implantation is broadly applied in semiconductor , and metallurgy technology [2,3] because of its advantage such as the low temperature process which allows to apply polymer masks in photolithography. Implantation is a kind of technology which allows introducing non-soluble elements into the material [4]. The process of irradiation is uniform in surface. It is easy to control the concentration of dopants introduced into depth. Thanks to the use of masks of various width, the concentration of dopants may be changed in various areas of surface in one cycle of implantation. The implantation process is well modeled with Monte-Carrlo simulation. Popular software for such modeling is SRIM code [5]. The experimental results agree very well with model results for Si, AIII-BV semiconductors as well as for metals. However, the parameters of implanted CdTe crystals differ considerably from those pointed by SRIM simulation. Therefore, the implantation of CdTE is less applicable in microelectronic technology of devices based on CdTe material. In this paper we present the problems encountered with ion implantation of CdTe. In Section 2 we present preparation of samples for implantation. The physical principles of implantation are discussed in Section 3 . The post implanted annealing of the samples is analysed in this section as well. In Section 4 the classical problems with implantation of B an As into CdTe in the aim of p-n junction formation are outlined. Structure damage of the samples after implantation is addressed in Section 5 basing mainly on the excellent paper of Rischau et al. [6].

Affiliations of authors should be typed in 9-point Times. They should be preceded by a numerical superscript corresponding to the same superscript after the name of the author concerned. Please ensure that affiliations are as full and complete as possible and include the country.

2 Preimplantation sample preparation

The majority of the samples were prepared in the form of platelets of a size of 5x4x1 mm^3 . Platelets were mechanically polished and chemically etched with a solution of bromine in methanol . The preimplantation annealing of Te rich CdTe in Cd vapour was used for 36 h at 500°C ensuring Cd overpressure of 0.5 Torr [7].

The implantation with As^+ at room temperature is not efficient while the damage of structure is so great that the sample becomes semi insulating . The recovery of crystal structure of the crystal structure was not achieved by post implanted annealing within two days at 650°C

[8]. Therefore, the hot temperature implantation had to be applied. For this , the obstacle of high vapour pressure of CdTe should be removed by encapsulating of the samples. It was done by means of SiO_2 layers .The layers with 150 nm are proper for As^+ ions implantation . SiO_2 encapsulating layers are stable for temperatures not exceeding 525° C [8].

3 Implantation overview, implantation parameters and instrumentation

The main implantation parameters are: temperature of target (T), energy of ions (E), ion current density (I), time of implantation (t), total dose (TD).

The typical set up for ion implantation consists of (Fig.1) source of ions, magnet for ion selection, accelerator, scanning electrostatic plates, beam diaphragm, The environment for all these parts/ factors is high vacuum (10^{-7} Torr).

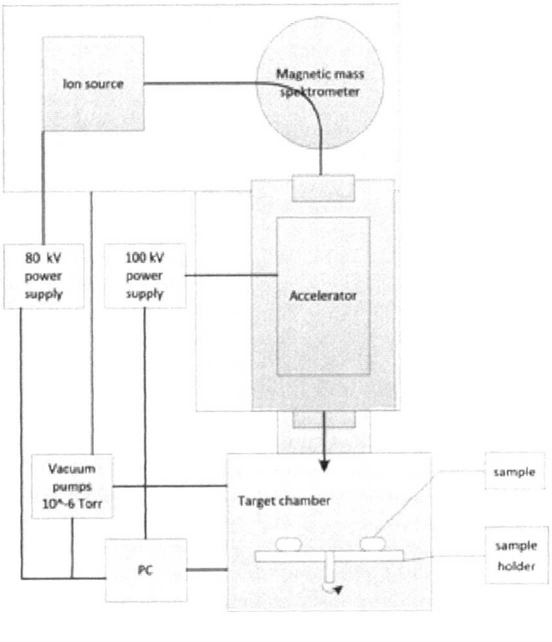

Fig. 1. Scheme of set up for ion implantation.

Ion current density is measured using Faraday box as:

$$j = I / q\,e\,A\,, \tag{1}$$

where A is the area of Faraday box , q is the iconicity number of ions (usually q =1), e is the elementary charge. The typical currents in ion implantation are 1 μA – 10 mA.

The fluency (dose) is defined as

$$\Phi = j \cdot t\ , \tag{2}$$

where t is the implantation time.

A dose gives a number of ions per unit area in whole depths of samples. It is not with simple relation to concentration of dopants (the number of atoms per unit volume while the concentration depends strongly on the depth although laterally it is constant.

Ions with high energy (usually 50 – 750keV) penetrates the near surface layer (up to 1μm for high energy). The ions undergo collisions with nucleons or electron of atoms of the target. These collisions are cascade and ions gradually lose the energy. Rate of energy loss into dept dE/dx depends on energy of ions and material. The dependence of the rate of energy loss for collision with electrons increases linearly versus $E^{1/2}$. Contrary to this, such dependence for collisions with nucleons is nonlinear showing clear maximum at energy in small range of energy E as it is seen in Fig. 2a . In this range of energy the nuclear loss of energy is dominated whereas for high energy ions the electronic loss is prevailing. In Fig. 2a and 2b Such rate of energy loss for CdTe crystals implanted with Cr, B, and As were calculated using SRIM code [5] and the results are demonstrated in Fig.2b and Fig.2c.

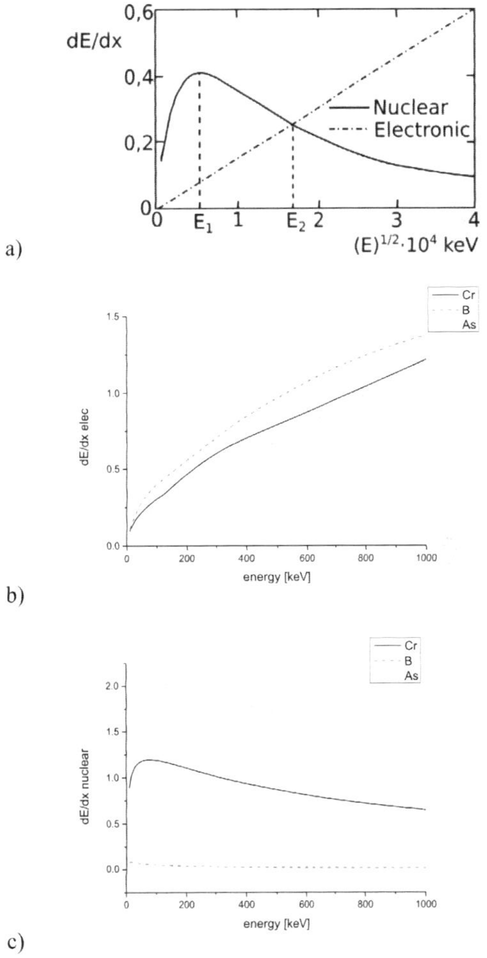

Fig. 2. Dependence of the rate of energy loss on $E^{1/2}$ (a), electronic (b) and nuclear (c) rate of energy loss for CdTe implanted with Cr,B and As (results of SRIM simulations).

DOI: 10.1051/epjconf/201713303002

Irradiation of a solid with high energy ions leads to numerous crystal defect formations on the surface as well in the near surface layer [9-12] :

1) surface defects are: sputtering , roughening, swelling, surface nanopatterning (see[13] and ref. therein),

2) in the bulk the following defects are created: vacancies, interstitials, pattern formation ,void formation, recombination, diffusion, sponge structure formation.

In the aim to remove these defects or part of them the annealing after implantation is applied. However, the annealing may create new defects like nucleation of existing defects, nanocluster formation, voids formation (by nucleation of vacancies) etc. [14-16].

The implantation of a small area on the surface of the target results in formation of bulk distribution of dopants in x,y, z space in the form of a ball (Fig .3a).

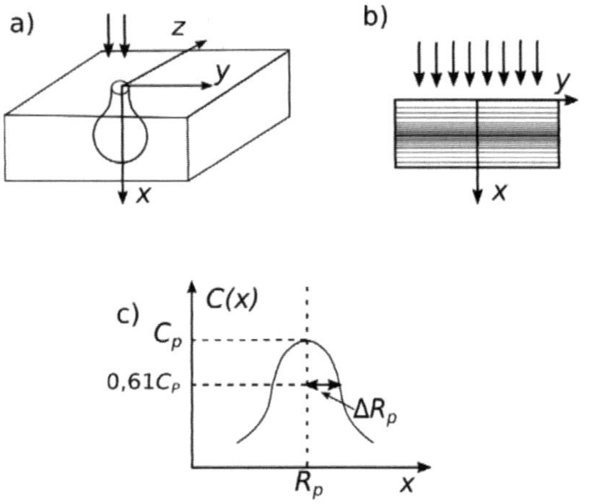

Fig. 3. Scheme of concentration distribution of defects (or dopants) in the sample implanted by stable ion beam (a), implanted by scanning ion beam (b), one-dimensional Gaussian approximation of defects distribution (c).

Usually the target is scanned by ion-beam , therefore the superposition of implantation at all the lateral position on a surface results in one-dimensional depth profile of dopant (Fig . 3b). Such profile can be approximated by Gaussian function (Fig. 3c):

$$C(x) = C_p \exp [- (x- R_p)^2 / 2(\Delta R_p)^2], \qquad (3)$$

where R_p is a projected range, and ΔR_p is a longitudinal straggle.

The parameters R_p and ΔR_p depend in a near linear way on graphs log R_p versus log E or log ΔR_p versus log E almost for all the elements implanted.

There are two aims of the post implanted annealing:

1) Restoring the crystallinity of the sample,

2) Introducing dopants into substitute sites of the crystal lattice in order to activate them(electrically or opticaly).

Three types of post implanted annealing are applied:

1) Steady annealing (e. g. T = 650° C , in time 1 hour for CdTe).

2) Rapid thermal annealing [17] using the xenon lamps (heating up to 1000°C with velocity more then 50°C/s).

3) Pulsed laser annealing with nanosecond length of pulses (see [18] and ref. therein).

4 The p-n junction formation in CdTe crystals

In the paper of Donnelly at al. [8] the following parameters were established with the aim of formation of type conversion of n - CdTe into p-CdTe by As^+ implantation : T = 500°C, E = 400 keV, I = 0.5 μA/ cm^2 , t = 15 min, dose = $3\cdot10^{15}/cm^2$. The p-n junction formation was supported by U-I characteristics which show forward resistance of 30 Ω and reverse breakdown of 40V.

Chu [7] implanted n-type undopted CdTe by As^+ ions with energy 60 keV, and dose 10^{15}. Electrical and photovoltaic properties were measured. Density of holes as a function of distance from implanted surface were determined by Hall measurements after periodic successive etching of the sample. From these measurements it follows that the effects of the implantation were observed at depts grater then 250 nm , whereas the modeling projected range for As^+ ions at such energy is about 29 ±17 nm.

Further continuation of implantation of CdTe crystals in order to obtain p- n^+ junction was ensured for p-type $Cd_x Hg_{1-x}Te$ [19,20]. The n- type conductivity may be produced by implantation of B [21], Al [22], In [23], Zn [23], Hg [24] as well as Au and P (see Capper [1] p.160). The electrical properties of ion implanted layers in $Cd_xHg_{1-x}Te$ were investigated by Margalit et al. [20]. Dopants (B, Ar, In, P, Au) were implanted at 150 keV, with dose 10^{11} -10^{15} 1/cm^2. The samples were annealed at temperature 150°C during 24 h in vacuum The samples were encapsulated by 200 nm thick CdTe layer. The results demonstrate that the crystal structure damage rather then kind of dopants converts p-type into n-type conductivity. The authors proposed to applicate the implantation for formation of accumulated surface for surface passivation and for formation of Ohmic contacts on n-type material.

Recently, ion implantation of CdHgTe is well known method for fabrication infrared photovoltaic devices. The paper [25] presents a short review on technology of production of CdHgTe fotodiodes by implantation As into n-type matrix.

5 Ion – beam induced damage in CdTe

Implantation damage in CdTe implanted with isotope of host ions ^{111}Cd at 60 keV at room temperature were investigated by Achtziger et al. [26]. Dose was low: $2\cdot10^{11}$1/cm^2 - $6\cdot10^{11}$1/cm^2 The host atoms were choosen for implantation in the aim to avoid any chemical activity of implants. Annealing treatment was carried out for several temperatures in the range 723K-1073K Characterization of the implanted and annealed samples was made by perturbed angular spectroscopy and by deep-level transient spectroscopy . For this last technique parallely, the set of nonimplanted samples was annealed and measured providing good reference data. The conclusions from these studies are: 1) At low doses all implanted atoms occupy substitutional lattice sites. Local crystal environment of these atoms is almost perfect before annealing:; after annealing at 673K it is perfect, 2) After implantation no additional electrically active centers have been observed.

C. W. Rischau et al. [6] have implanted CdTe single crystals with orientation (111) and (112). As dopants two types of ions were chosen: the light ions Ar^{+} and the heavy ions Sb^{++} having mass comparable with masses of self Cd and Te atoms. The implantation was carried out in two temperatures: 5 K and in room temperature. These two implantation temperatures should help to decide which mechanism of implanted defects formation, thermal or non-thermal is responsible for an unusually small, but deep damage of structure in CdTe. The energy of ions was 270 keV for Ar^{+} and 730 keV for Sb^{++}. Such values of energy were chosen to obtain the projected range for both ions $R_p = 220$ nm. Such value lies well beyond of the positions of surface atoms Cd and Te in RBS spectra. The dose was ranging from 10^{11} 1/cm^2 to 10^{17} 1/cm^2. The RBS spectra of samples irradiated with Ar^{+} ions did not differ considerably for both orientations of samples at room temperature. Moreover, at low temperature (15 K) this spectra are similar to those in room temperature . There is expressive dependence of lattice disorder on the dose . In spite of this, the channeling spectra do not reach random spectrum even for the highest dose $5 \cdot 10^{16}$ 1/cm^2 . This indicates that the crystal structure is resistive against amorphization even at a dose of 10^{17} 1/cm^2. The defect distribution obtained from RBS spectra show deformed Gaussian profiles with $R_p = 250$ nm independently on the orientation of crystals at room temperature . The tail of distribution is long and is equal to $z= 2R_p$ up to $z=4R_p$. This is a great discrepancy with SRIM results. The defects distribution at 15 K differs considerably from these latter. At doses higher than $2 \cdot 10^{15}$ the distribution is flat extended up to 4 R_p with the maximum at 350 nm independent on the orientation and type of ions. There is also significant increase of defects at surface of the sample. This surface damage increases significantly with the dose increase. However, for both temperatures the buried damage layer placed in the dept 100 nm below the surface and having the width of about 500 nm is formed, but this layer is not amorphous . Taking these results into consideration, Rischau [6] concluded that the high resistance of the CdTe crystals to amorphization follows not from thermal effects but from high ionicity of the crystal studied. Such conclusion were formulated earlier by Gettings at al. [27] and by Uzan-Saguy et al.[28].

Ion implantation of heavy Xe ions into (111) oriented single crystals of CdTe was studied by Petric at al. in the paper [29]. The disorder was investigated using Rutherford backscattering spectrometry (RBS) with He^{+} ions [30] . The simulation was done using the SRIM software [5].The energy of ions was 350 keV and the doses were as follows: $1 \cdot 10^{13}$1/cm^2 , $2 \cdot 10^{13}$1/cm^2 , $3 \cdot 10^{13}$1/cm^2 , $4 \cdot 10^{13}$1/cm^2 , $6 \cdot 10^{13}$1/cm^2 , $8 \cdot 10^{13}$1/cm^2 , $12 \cdot 10^{13}$1/cm^2 , $16 \cdot 10^{13}$1/cm^2. Parallely, for comparison, an identical implantation was done on Si single crystals. The experimental results from RBS measurements and results from modeling (SRIM) exhibit a great difference in full width at half maximum (FWHM) of damage profile created by Xe ions. For Si samples the FWHM is 150 nm whereas for CdTe samples the FWHM is 90 nm. These values are confirmed by RBS analysis. Moreover, RBS spectra made for Si and CdTe, both implanted with the Xe ions, show the great influence of disorder on the dose in the case of Si sample. For dose $6 \cdot 10^{14}$1/cm^2 the channeling spectrum is comparable with the spectrum of random RBS. Contrary to this, in the case of CdTe the channeling spectra for all doses lie near the virgin spectrum which means that the disorder is very weak even for doses exceeding amorphization threshold calculated from the SRIM code (several 10^{16} 1/cm^2). From the tails of RBS spectra follows that the extended defects , mainly dislocation loops, prevail in this material after implantation. This seems to be in conformity with results presented in papers of Leo et al. [31-33] where the formation of dislocation loops was explained by high mobility of interstitial Te and enhanced recombination of defects in the depth of the sample. On the other hand, the defects near the surface were grasped by the surface and therefore the near surface layer was protected again amorphization even for high doses.

6 CONCLUSIONS

Usually, the samples are encapsulated before implantation . It seems that this rule should be obligatory in the case of CdTe implantation with high energy. However, this problem is omitted in the works cited. The high stability of the crystal structure seems to promise auspicious future of this material. Its application in doping of new elements allows to obtain new properties of this material. Heavy doping with magnetic ions seems to be one of these prospective new application of ion implantation of CdTe crystals or layers.

Acknowledgments

The support from the Center for Innovation and Transfer of Natural Science and Engineering Knowledge at the University of Rzeszow is gratefully acknowledged.

References

1. P. Capper, *Properties of Narrow-Gap Cadmium-Based Compounds,* (United Kingdom 1994)
2. J. F. Ziegler (ed.) Ion Implantation (Science and Technology, Academic Press 1988)
3. E. Rimini, *Ion Implantation, Basic to Device Fabrication*, (Kluver Academic Publ., Boston 1995)
4. O. Meyer, A. Turos , Materials Science Reports **2**, 7, 373-468 (1987)
5. The Stopping and Range of Ions in Matter, http://www.srim.org.
6. C.W. Rischau, C.S. Schnohr, E. Wendler, W. Wesch,J. Appl. Phys. **109**, 113531 (2011)

7. M. Chu, A.L. Fahrenbruch, R.H. Bube, J.F. Gibbons, J. Appl. Phys. **49**, 322 (1978)
8. J.P. Donnelly, A.G. Foyt, E.D. Hinkley, W.T. Lindley, J.O. Dimmock, Appl. Phys. Lett. **12**, 303 (1968)
9. R. Böttger, K.H. Heinig, L. Bischoff, B. Liedke, R. Hübner, W. Pilz, Phys. Status Solidi - Rapid Res. Lett. **7**, 501 (2013)
10. R.A. Wilhelm, E. Gruber, R. Ritter, R. Heller, S. Facsko, F. Aumayr, Phys. Rev. Lett. **112**, 153201 (2014)
11. R. Böttger, K.H. Heinig, L. Bischoff, B. Liedke, S. Facsko, Appl. Phys. A Mater. Sci. Process. **113**, 59 (2013)
12. R. Ritter, R.A. Wilhelm, M. Stöger-Pollach, R. Heller, A. Mücklich, U. Werner, et al., Appl. Phys. Lett. **102**, 063112 (2013)
13. O. El-Atwani, S.A. Norris, K. Ludwig, S. Gonderman, J.P. Allain, Sci. Rep. **5**, 18207 (2015) 18207
14. S. Zhou, K. Potzger, G. Zhang, A. Mücklich, F. Eichhorn, N. Schell, et al., Phys. Rev.**B 75**, 085203 (2007)
15. S. Zhou, K. Potzger, J. Von Borany, R. Grötzschel, W. Skorupa, M. Helm, et al., Phys. Rev. **B 77**, 035209 (2008)
16. S. Zhou, A. Shalimov, K. Potzger, N.M. Jeutter, C. Baehtz, M. Helm, et al., Appl. Phys. Lett. **95**, 192505 (2009)
17. T.E. Seidel, D.J. Lischner, C.S. Pai, R.V. Knoell, D.M. Maher, D.C. Jacobson, Nuclear Instruments and Methods in Physics Research Section B: Beam Interactions with Materials and Atoms, 7–8, Part 1, 251-260 (1985)
18. S. Zhou, Dilute ferromagnetic semiconductors prepared by the combination of ion implantation with pulse laser melting (Topic Review), J. Phys. D: Appl. Phys. **48**, 263001 (2015).
19. C. Uzan-Saguy, J. Vac. Sci. Technol. A Vacuum, Surfaces, Film. **10**, 3246 (1992)
20. S. Margalit, Y. Nemirovsky, I. Rotstein, J. Appl. Phys. **50**, 6386 (1979)
21. M. Lanir, A.H.B. Vanderwyck, and C.C. Wang, J. Appl. Phys. **49**, 6182 (1978)
22. J. Marine and C. Motte, Appl. Phys. Lett. **23**, 450 (1973)
23. E. Igras, J. Piotrowski, and I.Z. Higersberger, Electron Technol. **10**,63 (1977)
24. G. Fiorito, G. Gasparrini, and F. Svelto, Appl. Phys. Lett. **17**, 105 (1978)
25. L. Mollard, G. Destefanis, N. Baier, J. Rothman, P. Ballet, J.P. Zanatta, et al., J. Electron. Mater. **38**,1805 (2009)
26. N Achtziger, J Bollmann, Th Licht, B. Reinhold, U Reislöhner, J Röhrich, M Rüb, M Wienecke, W Witthuhn, Semiconductor Science and Technology 11, 947 (1996)
27. M.Gettings, K.G. Stephens, Radiat. Eff. **22**, 53 (1974)
28. C. Uzan-Saguy, D. Comedi, V. Richter, R. Kalish, and R. Triboulet, J. Vac. Sci. Technol. A **7**, 2575 (1989)
29. P. Petrik, N.Q. Khánh, J. Li, J. Chen, R.W. Collins, M. Fried, et al., Phys. Status Solidi Curr. Top. Solid State Phys. **5**, 1358 (2008)
30. L.C. Feldman, J.W. Mayer, S.T. Picraux, *Materials Analysis by Ion Channeling*, (Academic Press 1982)
31. G. Leo, A. Traverse, M. O. Ruault, and A. V. Drigo, Nucl. Instrum. Methods Phys. Res. B **63**, 41 (1992)
32. G. Leo and M. O. Ruault, J. Appl. Phys. **73**, 2234 (1993)
33. G. Leo, A. V. Drigo, and A. Traverse, Mater. Sci. Eng. B **16**, 123 (1993)

DOI: 10.1051/epjconf/201713303003

Synthesis of dye-sensitized solar cells. Efficiency cells as a thickness of titanium dioxide

Dominika Szura*

Faculty of Mathematics and Natural Sciences, University of Rzeszow, Pigonia 1, 35-959 Rzeszow

Abstract. Defying the influence of the thickness of TiO_2 efficiency of dye-sensitized solar cell. It was confirmed that the compatibility of printed layers with the parameters closely related with the DSSC. It was found that the increase in thickness of the titanium dioxide layer, increases the distance between the electrodes, determined by the thickness of the Surlyn foil. With the rise of thickness of dyed layer of TiO_2 established decrease in the value of its transmittance. Greatest transparency and aesthetic value obtained for photovoltaic modules with a single layer of titanium dioxide. The improved performance efficiency and preferred yields maximum power were noticed and exhibited by the cells covered with three layers of TiO_2. It was established that the behaviour of economic efficiency in the production process, provides a range of cells with two layers of oxide, showing a similar performance and greater transparency.

1. Introduction

The development of technology and the continuous growth of population results in a dynamic demand for electricity. However access to conventional energy sources, which include coal, natural gas and oil is limited. It's worth pointing out, that their usage is associated with high levels of harmful emissions to the environment. For these reasons, in recent years they began to look for new solutions, paying great attention to alternative energy sources. Solar energy belongs to that group and it is used by the photovoltaic cells, converting it into electrical energy. History of dye solar cells dating back to 1887. However the breakthrough discovery was the year 1991. Gräetzel and his colleagues developed a cell structure of dye-sensitized solar cell. Low production cost and high efficiency caused that cells DSSC became the main object of research worldwide [1, 2]. The aim of the study was to investigate and determine the effect of influence of titanium dioxide thickness on efficiency of printed photovoltaic solar cells (DSSC).

2. Experimental Details

The first step was to cut glass sample using a diamond cutter. Prepared glass substrates were washed using ultrasonic cleaner in special detergents according to the procedure established in the photovoltaic Center Research and Development Company Ml System S.A. The next step was the ablation of the upper and lower glass made using a semiconductor laser of the fiber, in order to isolate each cell. Using a drill, holes were drilled located at the top and bottom edges of the glass. The layer is printed on the prepared glass substrate by screen printing. This is a technique in which the print form template is applied to the fine woven mesh, metal or made of synthetic fibers. Making a print is to paint by the swaging die. For samples without openings a titanium dioxide paste and silver paste, and the sample with holes platinum paste and silver paste were applied. In studies substrate glass was printed successively with one, two, three and four layers of the paste of TiO_2, in an effort to check if the thickness of the layer has influence on the efficiency of DSSC. The sample was subjected to a drying process in a drying tunnel each time. The purified stream of air samples were tested for the profilometer and then want to remove the organic material and combine grain active layers, were sintered processed at a temperature of about 480 ° C. After the heat treatment process surface topography was determined once again and profile layers of titanium dioxide. In a next step the immersion technique, the sample with printed layers of TiO_2 were stained in the solution of the dye N719 for 24h. Lamination process is carried out by combining a counter (representing the upper glass with holes) with photoanode (lower glass without holes) using Surlyn foil as an encapsulant. The final stage of the production was to fill the solar cells with electrolyte dye, based on a pair of redox I^- / I_3^- [3, 4].

*Corresponding author: dominika.szura@onet.pl

Fig. 1 Distribution of the samples in an industrial furnace in the sintering process.

Fig. 2 Laminated photovoltaic modules.

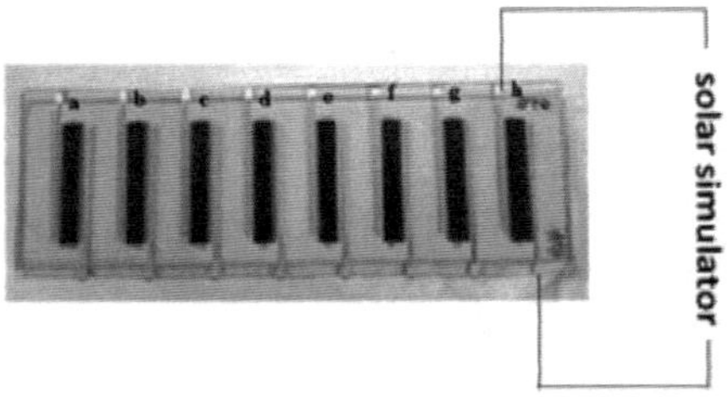

Fig. 3 Photovoltaic module with a description of individual cells and the way of connecting when removing the current-voltage characteristics.

The picture below presents a way of naming the individual cells (8), the photovoltaic module and method of removing the voltage signal.

3. Results and Discussion

The table shows the results of measurements of obtained TiO_2 layer thickness, before and after sintering process. Each module having one, two, three and four layers of titanium dioxide was measured on three cells. It was noticed that the thickness after firing decreased from 0.5-1 μm [5].

Table 1. The topography of the surface. The thickness of the layer of TiO_2, for photovoltaic modules with one, two, three and four layers of oxide.

THE TOPOGRAPHY OF SURFACE-THICKNESSES OF TiO_2 LAYER				
Name of cells	1 layer of TiO_2 [μm]	2 layers of TiO_2 [μm]	3 layers of TiO_2 [μm]	4 layers of TiO_2 [μm]
AFTER THE DRYING PROCESS				
a	5,5-6,0	9,0-10,0	15,0	20,0
d	5,5-6,0	9,0-9,5	14,0-14,5	20,5
g	6,0	8,5-9,0	15,0	21,0-22,0
AFTER THE FIRING PROCESS				
a	4,0-5,0	8,0-8,5	14,0	17,0-18,0
d	4,5-5,0	8,0-8,5	13,5-14,0	18,0
g	4,5-5,0	8,0	14,5	19,0-19,5

Using the SENSOMAP program made three-dimensional imaging topography surface and presented profile of the titanium dioxide layer. On its basis, it was found that layer is characterized by a much greater roughness before firing. After firing it shows a greater homogeneity.

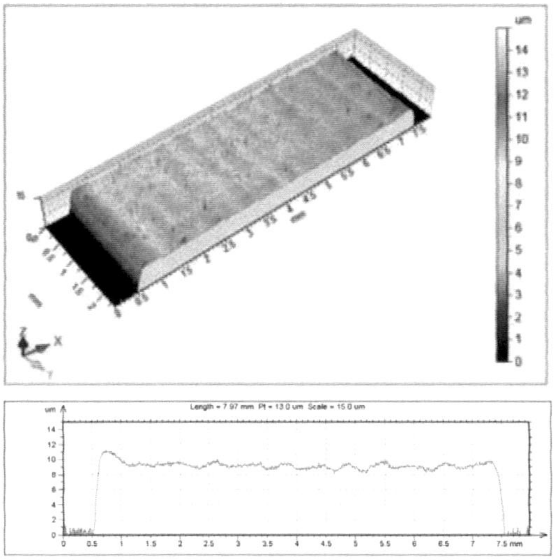

Fig. 4 The topography surface and profile of TiO2 glass layer substrate with two layers of oxide after the drying process.

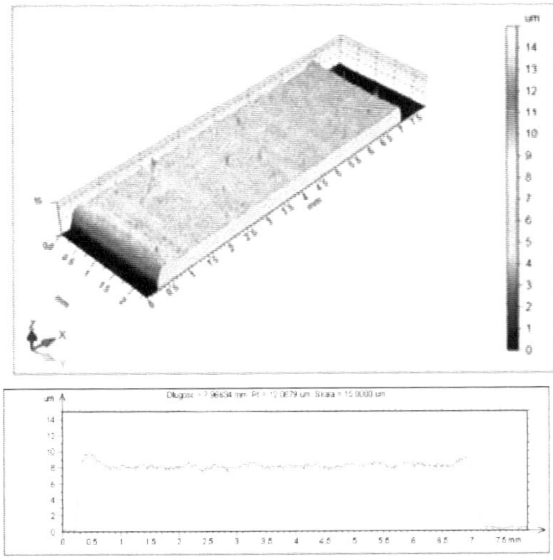

Fig. 5 The topography surface and profile of TiO2 glass layer substrate with two layers of oxide after the firing process.

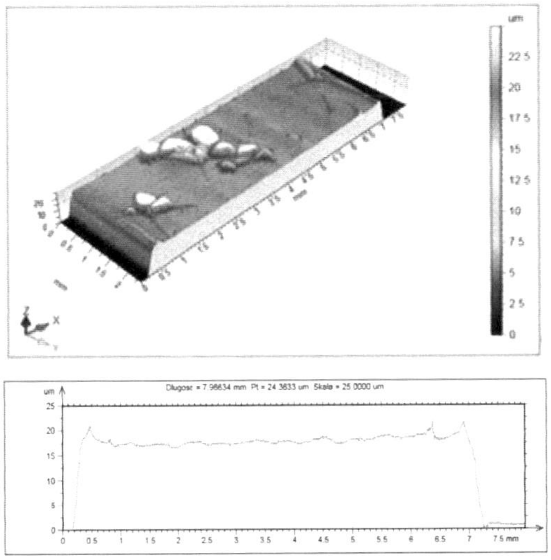

Fig. 6 The topography surface and profile of TiO2 glass layer substrate with four layers of oxide after the firing process.

The presence of defects in the case of photovoltaic cells with four layers of oxide was due to shrinkage of the fire and had a significant influence on the measuring results. In the high temperatures inside the industrial furnace, the water evaporated, organic matter degraded and particles of the raw material came closer to each other. Too much shrinkage caused mechanical deformation of the article [5].
Measurements of current-voltage characteristics were made using standard testing conditions (STC): (1000W / m^2, AM = 1.5, T = 25 ° C). The shape obtained characteristics in the case of cells of two or three layers of oxide it is virtually identical. For the remaining cells the chart starts with smaller current value. Based on the reports generated by special software simulator of the solar, containing current-voltage characteristics and electrical parameters for single cell of photovoltaic module, tables were prepared. The most important parameters were summarized there. In the recording accuracy of the median values, the value of the maximum probable error (d_x) using the methods of the Student-Fisher [6]. Obtained similar values of open circuit voltage and short circuit current for each cells of test modules with one, two and three layers of TiO_2 were obtained. In the case of cells of the four-layers oxide a value of short circuit current varied, and it was much lower. The highest maximum power equal to an average of 13.85 mW was obtained for a cell with three layers of titanium dioxide. This cell also showed the highest efficiency in the range of 3,76-4,04% [5].

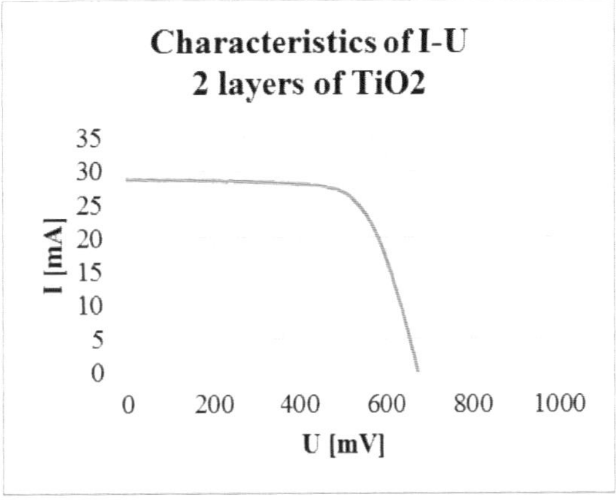

Fig. 7 Current-voltage characteristics for the cell of photovoltaic module with two layers of titanium dioxide.

Averaging other electrical parameters of particular cells of photovoltaic modules has been observed that increasing the thickness of the layer of TiO_2 causes a linear increase in short-circuit current (I_{sc}), maximum power (P_{max}) and efficiency (η). The results concerning the cell covered with four layers of titanium dioxide was excluded due to the inaccuracy caused by the cracking of the oxide layer during firing [5].

Table 2 Record the average values of individual parameters of photovoltaic cells, taking into the maximum probable error

Name of parameter	Test module- 1 layer of TiO_2	Test module- 2 layers of TiO_2	Test module- 3 layers of TiO_2	Test module- 4 layers of TiO_2
U_{oc} [mV]	677 ± 18	682 ± 22	673 ± 20	603,6± 7,1
I_{sc} [mA]	25,9 ± 1,2	28,2 ± 3,3	30,2 ± 2,2	5,0 ± 0,9
P_{max} [mW]	11,85 ± 0,31	13,0 ± 1,2	13,85 ± 0,72	1,95 ± 0,37
FF	0,68 ± 0,03	0,68 ± 0,02	0,682 ± 0,011	0,64 ± 0,03
η [%]	3,266± 0,081	3,57 ± 0,38	3,84± 0,22	0,53 ± 0,10

Based on the survey, it was found that the transmittance measurement of a dyed layer of TiO_2 decreases along with increasing of thickness. Transmittance indicates how much of the incident radiation is passed through materials. In the case of photovoltaic cells with three layers of TiO_2, it was observed that the light transmittance ranged from 8-10%. The highest value of transmittance of the order of 26-27% were obtained for the cells with one layer of oxide. It is worth noting that the decrease in the coefficient of transparency increases the efficiency of the module [5].

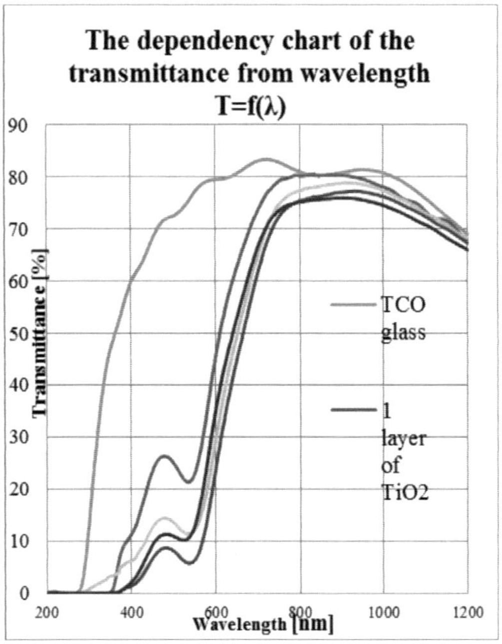

Fig. 8 Transmittance measurement for TCO glass and manufactured photovoltaic modules

4 Conclusions

In the manufacture of printed photovoltaic solar cells (DSSC) each stage of the production has a significant impact on the final result of product. We could list a lot of reasons such as correct cut glass substrate, the accuracy of laser ablation, drilling, proper selection of the thickness of the active layer, encapsulant, and screen printing process. Relating to a screen printing method, which was prepared photovoltaic modules with one, two, three and four layers of TiO_2, highlighted the rheological parameters of paste. Determining the quality of the products is the corresponding viscosity of the paste against to the size of the holes in sifter. In the case of too high viscosity of the paste which could occur in order in the printing inaccurate due to the lack overcome of flow resistance and the paste not connecting with each other. In addition, the thickness intensify of the reproducible relatively constant value was claimed. The topography of the surface of the active layer of TiO_2 solar cells was analyzed. The compliance of the printed layers with strict parameters regarding cells DSSC was confirmed. The cracking layer of titanium dioxide was noticed for cells with four layers of TiO_2 produced after firing. The presence of defects in the sintering was the result of contraction and had a significant impact on the results of measurements. Current-voltage characteristics was prepared. Comparable performance was observed for modules with the three and two layers of titanium dioxide, with a slight predominance of the former. It can therefore be concluded that with an increasing layer thickness of the titanium dioxide is advisable to use a larger distance between the electrodes which is assessed the thickness of Surlyn foil. Imposition of each layer of titanium dioxide is associated with additional costs connected with usage of a higher amount of dye or the electrolyte. In order to maintain the efficiency in the process, two layers of oxide could be used. Despite the better performance of efficiency and maximum power for cells with three layers. It was agreed that decreasing in the value of transmittance with a dyed layer of TiO_2 with increasing their thickness. In the case of photovoltaic cells with three layers of TiO_2 along, it was observed that the light transmittance ranged from 8-10%. The highest value of transmittance of the order of 26-27% were obtained for the cells with one layer of oxide. It is worth noting that the decrease in the coefficient of transparency increases the efficiency of the module [5]. To obtain photovoltaic cells characterized by greater transparency and aesthetic. It is recommended to choose the modules covered with a one layer of TiO_2. Taking into account the location and preferred yields power is better to use photovoltaic cells with three layers of TiO_2, showing the greatest efficiency. Referring to the economy it is worth to choose modules covered with two layers of TiO_2. They show a very similar performance with reference to the cells with three layers of titanium dioxide, moreover they are more transparent. However the final decision of the choice should belong to a potential buyer. In the Photovoltaic Centre for Research and Development still do research, which aim is to receive printed photovoltaic cells, in which it will be possible connection of performance, economy and aesthetics of products [5].

Aheartfelt thank you for Dr. hab. prof. UR Małgorzata Pociask-Biały for help, valuable advice and understanding when writing the work.

Thank you, Mr. Chairman Dawid Cycoń for the opportunity to undergo training in laboratory Photovoltaic Research and Development Center of ML System S.A. in which I had the opportunity to perform set of studies included in my thesis.

Mr. Dr. Michał Inglot, thank you for the tips, care and kindness during my stay in the laboratory of ML System S.A.

I offer my gratitude for the laboratory team workers of the Photovoltaic Research and Development Center, in particular: MSc. Natalia Trawka, MSc. Paweł Ciołkosz, MSc. Sławomir Chrobak, MSc. Michał Grudzień for the immense warmth, friendly working atmosphere, help in the synthesis of photovoltaic cells and valuable advice in mastering new techniques.

References

1. J.K. Lee, M. Yang, Materials Science and Engineering B, **176**, 1142– 1160 (2011)
2. S. Kim, M. K. Yeo, M. H. Um, M. Kang, Bull. Korean Chem. Soc. **33**, 4 (2012)
3. J. Szlachta, L. Marszałek, S. Chrobak, *Czysta Energia*, Abrys, **8** (2014)
4. J. Szlachta, L. Marszałek, S. Chrobak, E. Stanek, *Enkapsulacja ogniw DSSC warunkiem ich pomyślnej komercjalizacji*, Magazyn fotowoltaika, nr **4** (2014)
5. D. Szura, *Synteza fotowoltaicznego ogniwa barwnikowego. Wydajność ogniwa w funkcji grubości ditlenku tytanu.* Praca dyplomowa magisterska, Wydział Matematyczno-Przyrodniczy Uniwersytetu Rzeszowskiego, Rzeszów (2016)
6. H. Szydłowski, *Pracownia Fizyczna*, (Wydawnictwo Naukowe PWN, Warszawa, 1997)

Zinc oxide films prepared by spray pyrolysis

Andrzej Maciąg[1], *Piotr* Sagan[1], *Marian* Kuźma[1], *Volodymyr* Popovych[2]

[1] *Centre for Innovation and Transfer of Natural Sciences and Engineering Knowledge, University of Rzeszów, S. Pigonia 1, 35-959 Rzeszów, Poland*
[2]*Department of Fundamental Technologies, Ivan Franko Drogobych State Pedagogical University, 24 Ivan Franko str., 82100, Drogobych, Ukraine*

Abstract. In this work we developed and tested spray pyrolysis system for layers deposition. In the system we have used ultrasonic apparatus (nebulizator) as a sprayer. A zinc nitrate aqueous solution has been used as a precursor solution. The idea of the method is the decomposition of nitrate on a hot substrate according to the reaction $Zn(NO_3)_2 \rightarrow ZnO +2 NO_2 +1/2O_2$. The layers were grown on glass, (001)Si and KCl substrates at the temperatures 300 – 500^{0}C. The thickness of the obtained layers was in the range 50 – 500 nm, depending on the growth time and rate. The influence of substrate temperature on the morphology of the layers has been studied by SEM method. The energy gap of the layers was found to be the range of 3.26-3.3 eV from their absorption spectra.

1 Introduction

In the early 2000s T. Dietl and H. Ohno [1] predicted theoretically high temperature ferromagnetism in ZnO:Mn diluted magnetic semiconductor (DMS), which make it very prospective material for spintronic application. Basing on a Zener model of ferromagnetism [2], they calculated the value of Curie temperature in the range of 300 – 400 K for InN, GaN and also ZnO with 5% content of Mn and high hole concentration (more than $3.5{\cdot}10^{20}cm^{-3}$). Although after huge number of works with synthesis of such a materials the idea of room temperature DMS has not been verified experimentally for this material [3, 4], ZnO thin films have many other practical application as a material for variety of optoelectronic and electroacoustic devices (see *e.g.* the review paper [5]). In numerous laboratory ZnO thin films were deposited by well-known techniques, such as molecular been epitaxy [6, 7], magnetron sputtering [8], chemical vapor deposition [9], pulsed-laser deposition [10] and atomic layer deposition [11, 12]. In this paper we have applied spray pyrolysis method, which is more rarely used for the sake of thin layers or nanoparticles fabrication. Normally, this method are applied for fabrication of thin-film gas sensors (*e.g.* SnO_2) and conducting and transparent ZnO:In layers [5], as well as for carbon coating of electrodes in fuel cells technology and in lithium-ion batteries production. The main advantages of the spray pyrolysis are small cost of the set up, flexibility of the process, low temperature and vacuum less processing, uniform layer coating, possibility of coating of large area substrates. The aim of present paper is to prepare ZnO thin layers by pyrolysis method [13, 14] and test their structural and optical quality. For our purposes, we have developed a homemade system for small (5×5 mm^2) and large (10×10 cm^2) area layer deposition. We have obtained ZnO layers using $Zn(NO_3)_2$ as a precursor. For structural measurements the layers were grown on Si substrates, whereas glass substrates were used for optical and electrical characterization of the grown layers.

2 Experimental

2.1. A homemade system for spray pyrolysis

A homemade apparatus for spray pyrolysis (Fig. 1) consists of ultrasonic nebulizer, nozzle, and heating plate. An air was used as a carrier gas. Ultrasonic nebulizer is connected with air pump and with nozzle mounted on X–Y throat. The position and movement of the nozzle in X-Y plane was controlled by computer. Deposition was carried out via scanning the surface of the substrate by nozzle in 20 rows, each of 3 mm wide. The flow of air gas was controlled by flow meter. Heating plate was equipped with contact thermometer and temperature controller. The temperature was changed in the range of 100 – 600°C. Single crystalline (100)-oriented Si, glass and single crystalline KCl were used as substrates. The deposition of the layers was carried out at various technological parameters:
1. The distance of nozzle – substrate: 1, 2, 3 and 4 cm; 2. Flow of air: 3.3 min^{-1}; 3. Temperature of substrate: 200 – 600^{0}C; 4. Concentration of precursor in water: 0.1 M.
All samples presented in this paper were deposited onto substrates heated to the 600 ^{0}C, at precursor concentration of 0.1 M and nozzle-substrate distance of 20 mm.

Corresponding author: psagan@ur.eu.pl

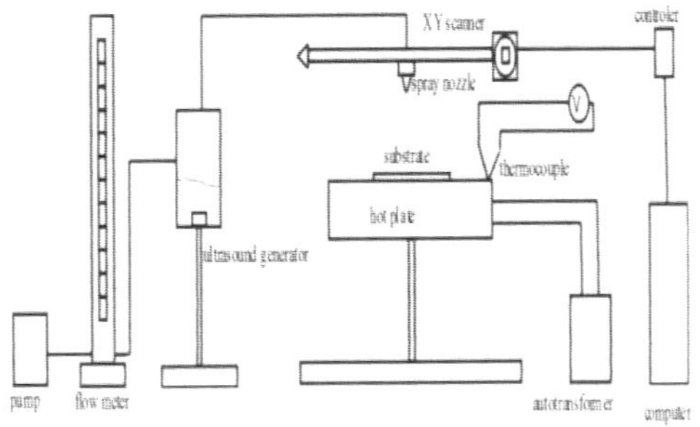

Fig. 1. Spray pyrolysis system.

Zinc nitrates appear in the form of four hydrated and one anhydrous (Table 1).

Table 1. Forms of zinc nitrates and their melting temperatures.

Form of $Zn(NO_3)_2$	Melting point [K]
$Zn(NO_3)_2{\cdot}H_2O$	343,9
$Zn(NO_3)_2{\cdot}2H_2O$	328,6
$Zn(NO_3)_2{\cdot}4H_2O$	317,9
$Zn(NO_3)_2{\cdot}6H_2O$	308,3
$Zn(NO_3)_2$	623 (decomposition temperature)

Thermal decomposition of zinc nitrates was extensively studied by B. Malecka [15] by using of mass spectroscopy technique. The decomposition is a many stage chemical process via intermediate products. In our experiment we were working with heksahydrate $Zn(NO_3)_2{\cdot}6H_2O$. At 338 K an intermediate compound $Zn(NO_3)_2{\cdot}Zn(OH)_2{\cdot}2H_2O$ are formed during slow thermal decomposition. At 393 K another compound, $Zn(NO_3)_2{\cdot}2\,Zn(OH)_2$, is formed. In general, the kind of the intermediate phase depends on the temperature and heating rate [15]. At temperature greater than 250°C the metastable phases transform into ZnO. Besides of ZnO, oxygen and nitrogen oxides are the final products. If the temperature of pyrolysis was too low, the residual of nitrogen intermediate products were observed in the layers by registering their different colors. Scheme of the thermal decomposition of $Zn(NO_3)_2\,{\cdot}6H_2O$ can be written as [16]:

$$Zn(NO_3)_2 \rightarrow \left\{ \begin{matrix} Zn(OH)_2 \\ Zn_2O(NO_3)_2 \\ Zn(OH)(NO_3) \end{matrix} \right\} \rightarrow ZnO + yN_wO_x + zO_2 \quad (1)$$

2.2. Characterization of the layers

Structure of the layers, deposited onto glass substrates, was investigated using optical microscopy. The films, grown on Si and glass substrates, were studied by means of scanning electron microscopy (SEM) using Vega 3 Tescan electron microscopy (U = 25 kV). The composition of the layers was measured by energy dispersive X-ray (EDS) method on this electron microscopy equipped with Bruker Quatax software and 10 mm^2 area, 126 eV resolution detector. The measurements absorption spectra for determination of the layers' energy gap were carried out on Carry UV-3100 spectrometer in the wavelength range of 200 – 1200 nm.

3 Results and discussion

Solid zinc nitrate begin to decompose thermally at 350°C. However, the start of spray pyrolysis with aqueous solution of zinc nitrate was observed at lower temperatures 150 – 170^{0}C. Studenikin *et al.* [16] showed that the layers obtained in such low temperatures are of bad quality. The critical temperature of 180°C has been introduced, above this temperature (at 180 – 400°C) the layers are very smooth to the eye and almost transparent. In Table 2 the preparation conditions of the investigated layers are collected.

Fig. 2 presents SEM images of the ZnO layers obtained on the chemo-mechanically polished and mirror-like (100)-oriented surface of single crystalline Si. The layers were deposited at 300°C, 400°C and 500°C. Low magnification images (µm-scaled) of the layers are presented on the left panel of Fig. 2, while the images in a greater magnification are collected on the right panel. Photos on the right indicate good morphology of the proper layers, without drops visible on the left images. The layers contain drops with diameter of about 2 µm (Fig. 2 a, c, e), which is the same for each temperature of pyrolysis. The value of the drop on the films surfaces corresponds well with that of drops in aqueous solution of zinc nitrate in aerosol. However, the morphology of the drops changes with the increase of the growth temperature. At 300°C drops are of regular form, at 400°C some drops have holes inside (Fig. 2c) and at 500°C a number of drops are in the form of the rings with well-defined outside faces (Fig. 2e). Such a change of the drops morphology reflects the corresponding steps of process evolution during the thermal decomposition of the solution of the precursor. The sequence of the decomposition is as follows (Fig. 3): atomization → solvent evaporation by diffusion → shell formation →evaporation by open shells → decomposition of particles or shells → crystallization.

The rate of evolution presented in Fig. 3 depends on time of the thermal treatment of the particles and on temperature of the substrate. The size of the particles also influences the morphology of final product. In the scanning method of layers formation the time of deposition is much shorter in respect to the time of "thermal annealing" during deposition and after deposition (cooling of the sample). Therefore, we relate the change of morphology with the annealing of the samples. In Fig. 2a one can notice solid particles with very rough shell. It means that at this temperature the process of evaporation was not finished. At growth temperature 400°C solid shells are visible and some of them are damaged and empty inside. It is clear from Fig. 2e that the temperature 500°C was high enough in order that all drops become hollow.

Most of the shells are crushed forming flat layer, which structure is not observable by SEM microscope even at considerably high magnification (Fig. 2f).

Corresponding author: psagan@ur.eu.pl

 DOI: 10.1051/epjconf/201713303004

Table 2. Technological parameters of the ZnO thin films, chosen for the studies (C is a concentration of precursor, D is a spray nozzle – substrate distance, t_p is a time of pyrolysis).

Sample No.	T (°C)	Substrate	Thickness [nm]	C (M)	D (mm)	t_p (s)	Results
39	300	Si(001)	40	0.1	30±1	600±1	SEM
52	400	Si(001)	80	0.1	20±1	600±1	SEM
41	500	Si(001)	70	0.1	20±1	600±1	SEM
36	300	glass	40	0.1	20±10	600±1	SEM
53	300	glass	40	0.1	20±1	600±1	R(t)

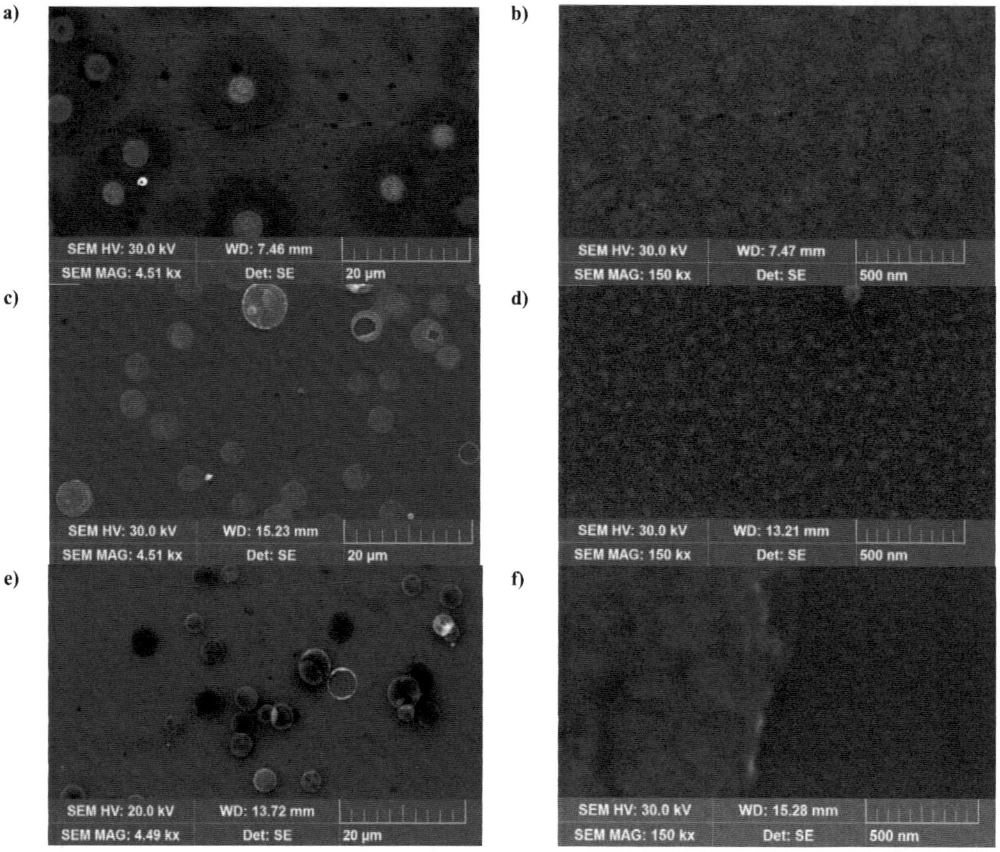

Fig. 2: SEM images of the samples No. 39 (a, b), No. 52 (d, e) and No. 41 (e, f) obtained at temperatures 300^0C, 400^0C and 500^0C, respectively.

Corresponding author: psagan@ur.eu.pl

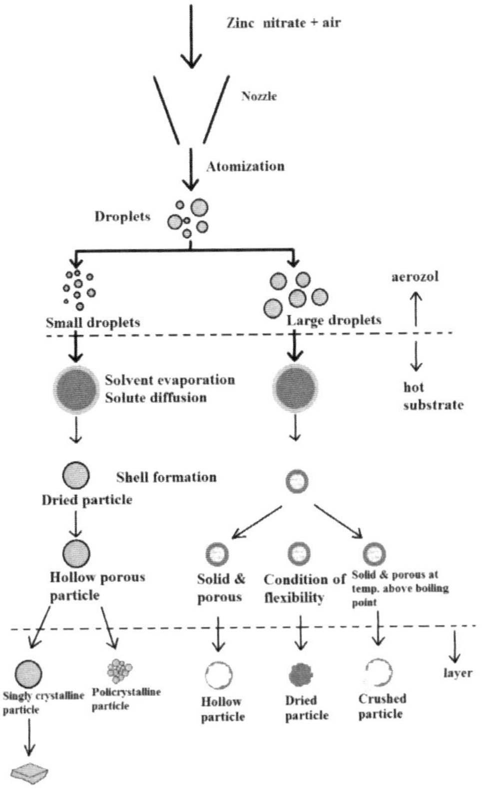

Fig. 3. Evolution of decomposition of the droplets in spray pyrolysis method (reproduced from Fig. 1 of paper [13]).

In Fig. 2c single particles (dark balls) are visible as well. The structure of these balls was not investigated in the present paper. The proper layer obtained at 300°C (Fig. 2b) has fractal structure [17]. It consists from aggregates (with diameter about 500 nm) built from the small individual 10 nm-sized particles. These later are composed from the smaller ones, *etc.* Such morphology is not observed for the layers obtained at 400^{0}C and at 500^{0}C (see Fig. 2 d, f). Chemical composition of layers measured by EDS method points on good stoichiometry 1:1. [Fig.4]. Due to a small thickness of the layers (tens nanometers) the EDS accuracy of the measure is not high. Therefore, the layer measured was spited from the substrate and composition of obtained powder of ZnO was determined.

The results EDX measurements, shown in Fig. 4, indicated deviation of the stoichiometry of the obtained films to oxygen-enriched side. But exact determination of their composition is hindered by the fact that generation region of characteristic X-rays in ZnO in EDX method is higher than thickness (tens of nanometers) of the layers. Therefore, substrates could contribute to the recorded EDX signal, lowering the accuracy of the analysis.

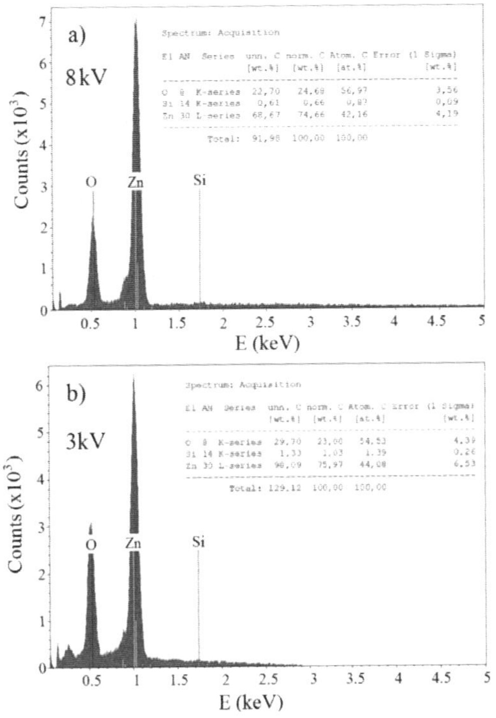

Fig. 4. Composition of the layers No. 36, measured by EDX at the applied voltages of 8 kV (a) and 3 kV (b).

Transmission spectrum of the sample No. 53 (Fig. 5a) revealed very sharp absorption edge, which are typical for most of the layers obtained in this work. In Fig. 5b spectral dependence in the narrower interval near the fundamental absorption edge is presented. Although the dependence is not exactly linear at this energy range, we determined the energy gap E_g using formulae for the direct interband transitions:

$$\alpha \cdot h\nu = C(h\nu - E_g)^{\frac{1}{2}} \qquad (2)$$

where *hv* is a photon energy. We obtained the values in the range 3.24 – 3.30 eV for the several measured samples, which are less values compared to those (3.3 eV) reported by Studenikin *et al.* [16]. We did not notice any dependence of the energy gap on the temperature of pyrolysis.

All samples were of high electrical resistance (more than 1 MΩ), as it was determined by two-contact method. Their photoconductive relaxation was tested using mercury lamp equipped with ultraviolet filter. Measurements of transient photoconductivity after switching of the lamp revealed long (tens of minutes) relaxation time (Fig. 6), the exact nature of which is not clear yet.

Corresponding author: psagan@ur.eu.pl

DOI: 10.1051/epjconf/201713303004

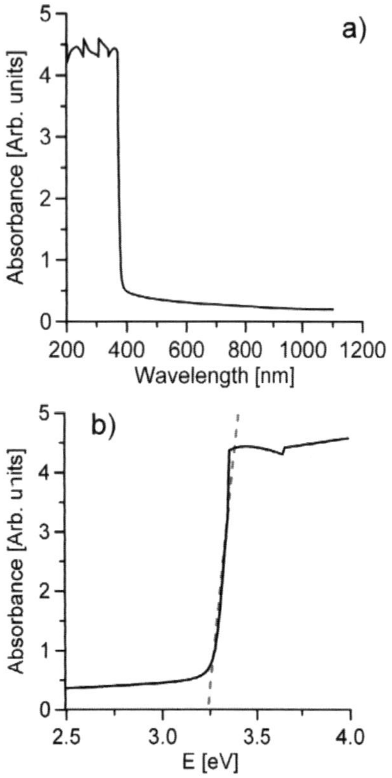

Fig.5: Absorption spectrum of the sample No. 36.

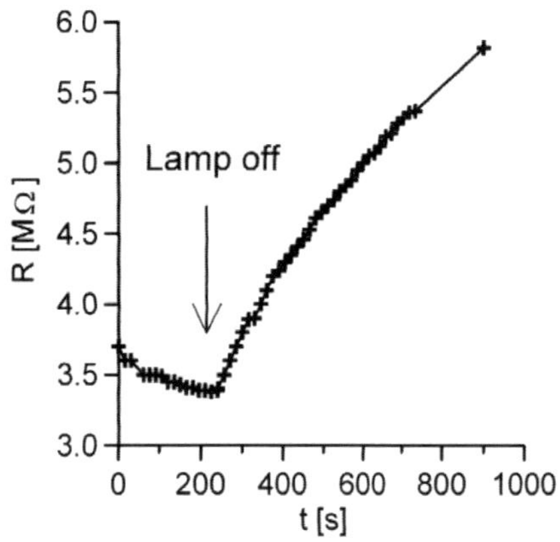

Fig.6: Relaxation of photoresistance for the sample No. 53.

4. Conclusions

The ultrasound spray pyrolysis is an advantage technique for thin films fabrication. It is unexpensive and vacuum-less method suitable for mass production of selected semiconductor films, such as ZnO and CuO. These materials are good candidates for photovoltaic industry due to the low cost of Zn and Cu metals. In this paper we presented a simple homemade system for pyrolysis, which is proved to be suited for fabrication of both small and large-area (up to 10×10 cm^2) ZnO thin films. The morphology of the deposited films indicated their polycrystalline character regardless of the structure of the used substrate (single crystalline (100)-oriented Si or amorphous glass). The size of crystallites in the deposited layer can be controlled by tuning the size of droplets in the spray, which is controlled by the parameters of the spray generator. SEM studies of the layers obtained at 300°C, 400°C and 500°C confirmed the model of evolution of morphology of zinc nitrate drops during the pyrolysis process, previously published by K. Okuyama et al. [13]. The deposited layers demonstrated good optical quality and crystallinity, being almost transparent in the visible region of the spectrum, and are characterized by sharp absorption edge (having small tail only), which is not dependent on the technological parameters of layers fabrication. The obtained layers are of high electrical resistance (in order of MΩ) and show longtime (up to tens of minutes) characteristic relaxation of photoelectrical response.

Corresponding author: psagan@ur.eu.pl

Acknowledgments

This work was supported by Centre for Innovation and Transfer of Natural Science and Engineering Knowledge at the University of Rzeszow (Poland).

References

1. T. Dietl, H. Ohno, F. Matsukura, Phys. Rev. B, **63**,195205, (2001)
2. C. Zener, Phys. Rev. **81**, 440 (1950)
3. T. Dietl, Nature Mater. **9**, 965-974 (2010)
4. S. J. Pearton, C. R. Abernathy, M. E. Overberg, G. T. Thaler, D. P. Norton, N. Theodoropoulou, A. F. Hebard, Y. D. Park, F. Ren, J. Kim and L. A. Boatner, J. Appl. Phys., **1**, 93 (2003)
5. Ü. Özgür, Ya. I. Alivov, C. Liu, A. Teke, M. A. Reshchikov, S. Doğan, V. Avrutin, S.-J. Cho and H. Morkoç, J. Appl. Phys. **98**, 041301 (2005)
6. Zhengwu Jin, T. Fukumura, M. Kawasaki, K. Ando, H. Saito, T. Sekiguchi, Y. Z. Yoo, M. Murakami, Y. Matsumoto, T. Hasegawa, and H. Koinuma, Applied Physics Letters **78**, 3824 (2001)
7. David C. Look, D. C. Reynolds, C. W. Litton, R. L. Jones, D. B. Eason, and G. Cantwell, Appl. Phys. Lett., **81**, No. 10, 1830 (2002)
8. Kachirayil J. Saji, Sascha Populoh, Ayodhya N. Tiwari and Y. E. Romanyuk, Phys. Stat. Sol. A **210**, 1386–1391(2013)
9. An-Jen Cheng, Yonhua Tzeng, Yi Zhou, Minseo Park, Tsung-hsueh Wu, Curtis Shannon, Dake Wang, Applied Physics Letters **92**, 092113 (2008)
10. Yolanda Y. Villanueva, Da-Ren Liu, Pei Tzu Cheng, Thin Solid Films **501**, 366 – 369 (2006)

11. E. Guziewicz, M. Godlewski, T. Krajewski, Ł. Wachnicki, A. Szczepanik, K. Kopalko, A. Wójcik-Głodowska, E. Przeździecka, W. Paszkowicz, E. Łusakowska, P. Kruszewski, N. Huby, G. Tallarida and S. Ferrari , J. Appl. Phys. **105**, 122413 (2009)

12. T.A. Krajewski, G. Luka, P.S. Smertenko, A.J. Zakrzewski, K. Dybko, R. Jakiela, L. Wachnicki, S. Gieraltowska, B.S. Witkowski, M. Godlewski and E. Guziewicz, Acta Physica Polonica **120**, 17-21 (2011)

13. K. Okuyama, I. Wuled Lenggoro, Chemical Engineering Science **58**, 537–547 (2003)

14. L. Filipovic, S. Selberherr, G. C. Mutinati, E. Brunet, S. Steinhauer, A. Köck, J. Teva, J. Kraft, J. Siegert, F. Schrank, Proceedings of the World Congress on Engineering 2013, Vol. II, WCE 2013, 3 – 5 July, London, U.K (2013)

15. B. Małecka, R. Gajerski, A. Małecki, M. Wierzbicka, P. Olszewski, Thermochimica Acta **404**, 125–132 (2003)

16. S. A. Studenikin, N. Golego, M. Cocivera, J. Appl. Phys. **83**, 2104 (1998)

17. G. Wisz, T. Ya. Gorbach, P. S. Smertenko , A. Błahut, K. Zembrowska, M. Kuzma, Superlattices and Microstructures **36**, 353-358 (2004)

Corresponding author: psagan@ur.eu.pl

DOI: 10.1051/epjconf/201713303005

Transmittance of selected nanostructurized solar glasses designated via relative change in electrical parameters of silicon solar cells

Malgorzata Pociask-Bialy*, *Kornelia* Kalwas

Faculty of Mathematics and Natural Sciences, University of Rzeszow
Centre for Innovation and Transfer of Natural Sciences and Engineering Knowledge,
S.Pigonia 1, 35-959 Rzeszow, Poland

Abstract. Photovoltaics is one of the most promising technologies for electricity production. In the future, photovoltaics could be an effective and safe source of energy.
In this work were present the results of the analysis of a special solar glasses transmissivity coefficient used as protective cover of photovoltaic cell. Antireflective glass due to its unique physical properties eliminate reflections and significantly increasing light transmission. The study of the relative change in the electrical parameters of photovoltaic cells ,with and without coats, as open-circuit current I_{SC} and the maximum power point MPP are presented in this paper. Research were undertaken with using the solar simulator QuickSun130CA, Class AAA+, under Standard Test Conditions.

1 Introduction

Thanks to the use of nanotechnology and the introduction of morphological changes of the glass surface used in the manufacture of solar glasses, D.A. Glass Company, were achieved extremely high performance light transmittance of samples of solar glasses measured at STC. We able to design transmittance of these solar glasses with use of the solar cell simulator QS130CA, Endeas Oy, Finland.

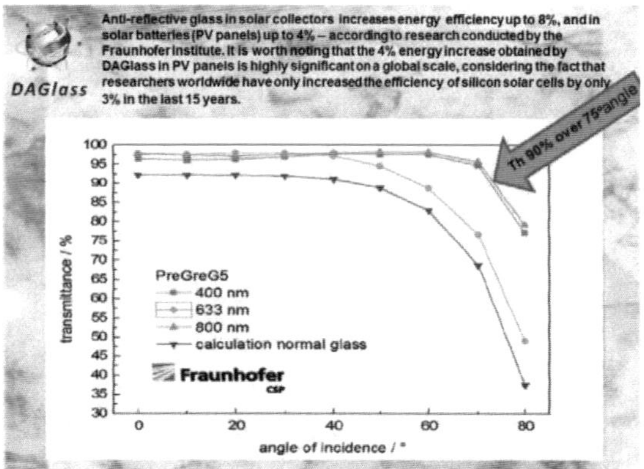

Fig. 1. Increasion of transmittance of DAGlass103 type of glass measured for the angle of incident ray over 75^0 [1].

D.A. Glass Company products were tested. T_d ratio is 99.3%. The aim performed tests was provided to check the transmissivity of solar glasses samples manufactured at D.A. Glass, which the company set and published on its website. The specific distinguishing feature of the market glasses solar company D.A. Glass is to improve permeability in the range 700 – 900 nm .
A special product of the company are NSTM AR glasses. Samples are performed via chemical and nanostructurised etching provided to inbuilding into surface structure of solar glasses the micro inclusions similar to short spaghetti. The visible increasing of transmittance DAGlass103 type of glass is registered for the angle of incident ray over 75^0 (see Fig. 1).

2 Samples and methods

There were tested six samples of solar glasses, D.A. Glass Company. All measurements were performed on QS130CA solar simulator at Renevable Energy Sources Laboratory of Centre for Innovation and Transfer of Natural Sciences and Engineering Knowledge of University of Rzeszow.
Before calculation we must describe the impact of glass area in the degradation of I_{SC} and P_{MM} (Fig. 2), the selected parameters of solar cell,when we applied to a cell a solar glass sample. There were measured a group of six different samples. The way how the measurements were performed is presented on Fig. 2.
Using an optical profilometer ContourGT, Bruker (Fig. 3) there were examined the structure of anti-reflective coatings.

* Corresponding author: pociask@ur.edu.pl

DOI: 10.1051/epjconf/201713303005

Fig. 2. Solar glass sample placed on referency solar cell. The signal is recorded in 45,2% from the cell masked by glass, and in 54,8% from the non masked cell.

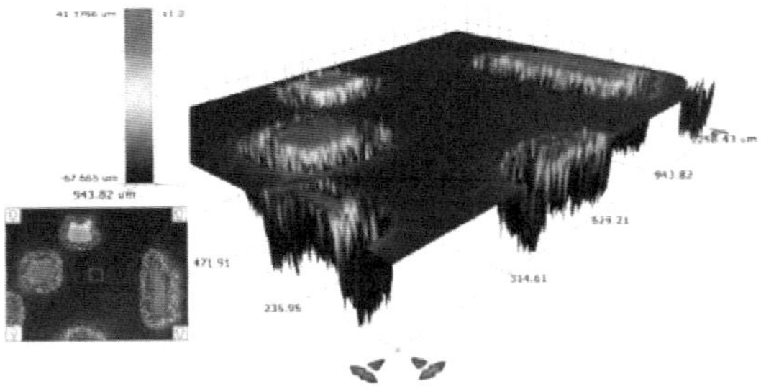

Fig. 3. Results of profile measurements with help of ContourGT, an example for sample No4.

The mean difference between the peak asperities, and the smooth surface of the glass were measured.To measure the density of surface roughness for all samples has been used a *Pro HR* Dino-Lite microscope. The measured and calculated parameters of all investigated probes are listed in Table 1. Sample No6 is lack of nanostructurized inclusion.

Table 1. Parameters of solar glasses, specially S_i/S – the real surface of solar cell from which the *I=f(U)* signal was measured.

Sample No	Sample image, Profile	Profile high [μm] of antireflection coat via X-axis and Y-axis	Sample image	Area, S [cm²]	S/S	Depth, d [mm]
Referency Sample	c-Si solar cell		c-Si solar cell	243,36	45,20	
1		X: 27,2304 Y: 26,4061	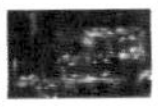	110,01	44,89	3,82
2		X: 29,1359 Y: 29,7936	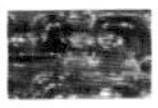	109,24	44,89	3,82
3		X: 30,8609 Y: 33,5099		108,38	44,53	3,82
4	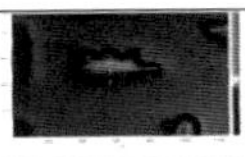	X: 15,4092 Y: 14,7752		108,99	44,78	3,84
5		X: 29,6276 Y: 31,3585		107,06	44,00	3,81
6				101,09	41,54	3,82

There were measured typical optical parameters of solar glasses like transmittivity *T* and absorption *A*.

Transmitivity *T*:

$$T_I = \frac{I_{scR}}{I_{sc(ref)}} \cdot 100\%, \qquad T_P = \frac{P_{MPR}}{P_{MP(ref)}} \cdot 100\% \tag{1}$$

Real value of open current, I_{scR}:

$$I_{scR} = \left(I_{sc\,(1,2,3,..)} * \frac{S_{(1,2,3,..)}}{S_{(ref)}} \right) + \left(I_{sc\,(ref)} * [1 - \frac{S_{(1,2,3,..)}}{S_{(ref)}}] \right) \tag{2}$$

Real value of maximum power, P_{MPR}:

$$P_{MPR} = \left(P_{MP\,(1,2,3,..)} * \frac{S_{(1,2,3,..)}}{S_{(ref)}} \right) + \left(P_{MP\,(ref)} * [1 - \frac{S_{(1,2,3,..)}}{S_{(ref)}}] \right) \tag{3}$$

And transmittancy *T* described throught both current (fotmula (2)) and power (formula (3)) parameters, as a final result :

$$T = T \pm |T_I - T_P|_{\max} \tag{4}$$

The absorption coefficient of solar glasses is done clasically:

$$I_{scR} = I_{sc(ref)} e^{-Ad} \tag{5}$$

where *d* – depth of solar glass,
A – absorption coefficient of solar glass.

$$\frac{I_{scR}}{I_{sc(ref)}} = e^{-Ad}, \tag{6}$$

and optical density *E* is given by:

$$log \frac{I_{sc(ref)}}{I_{scR}} = E, \tag{7}$$

$$E = Ad \cdot log \cdot e = 0{,}4343 \cdot Ad \tag{8}$$

$$A = \frac{E}{0{,}4343 \cdot d} \tag{9}$$

and

$$R + A + T = 1. \tag{10}$$

Our investigations were focused on describing transmittancy T, and reflectivity coefficient R was only calculated withot any measurements. Our goal was to measure transmittancy T.

The silicon cell operates much worse overlaid by a solar glass. Such a impediment placed on the road incident on cell radiation significantly reduces its performance. Electrical parameters of such setup (Fig. 2) were measured with help of solar simulator QS130CA.

* Corresponding author: pociask@ur.edu.pl

DOI: 10.1051/epjconf/201713303005

Fig. 4. QuickSun 130CA solar simulator, Endeas Oy, Finland [2].

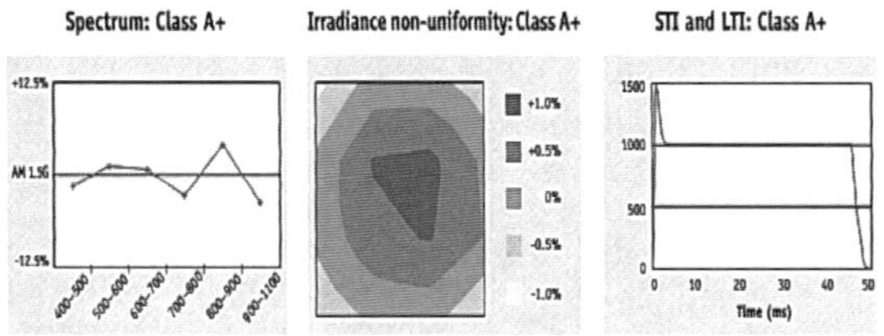

Fig. 5. QuickSun 130CA solar simulator, Endeas Oy, Finland class AAA+ [2].

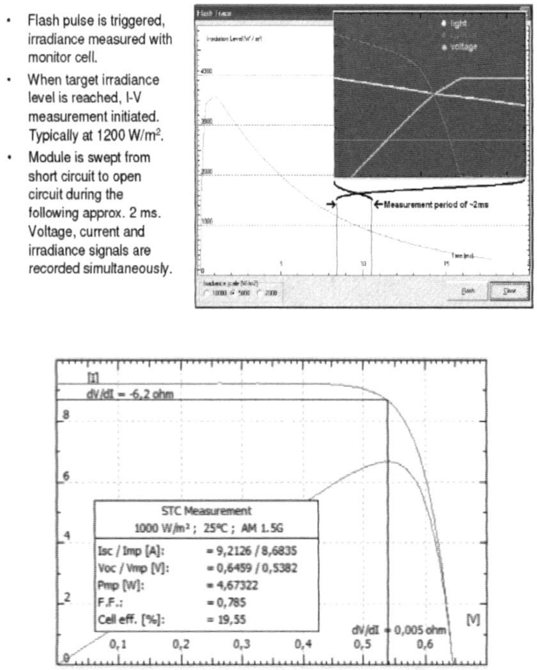

Fig. 6. QuickSun130CA Software. The idea of measurements in brief [2].

* Corresponding author: pociask@ur.edu.pl

3 Results and discussion

There were measured both sides of investigated samples: smooth and nanostructurized. Every test was repeated twice, and average value represents a final result of measurements performed on QS130CA solar simulator [3].

In Fig. 7 are presented selected results of tests for sample No 1. There were three curves, each indicates voltage-ampere characteristics: black – reference cell signal, red – signal taken from cell and glass from nanostructucturized side, and green - signal taken from cell and glass from smooth side.

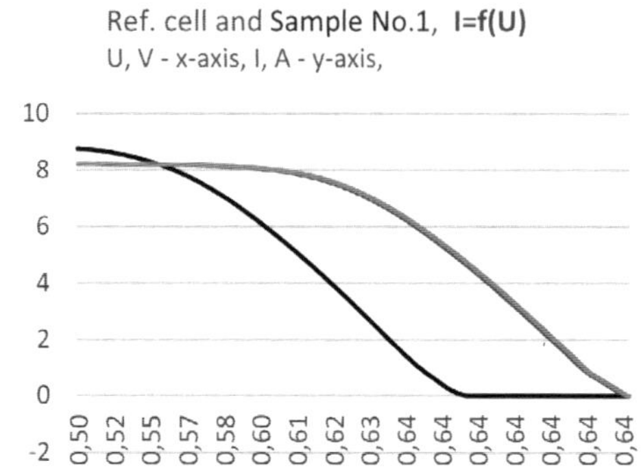

Fig. 7. Results of tests done with help of QS130CA for sample No 1 with certificate of measurements.

STC: 1000W/m^2; 25°C; AM 1.5G		Sample No. 1	
Parameter	Ref. cell	nanostruct.	smooth
Isc [A]	8,945	8,254	8,253
Voc [V]	0,643	0,642	0,642
Pmp [W]	4,55	4,19	4,20
F.F. [%]	79,063	79,173	79,16
Cell eff. [%]	18,7	17,233	17,237

In Tables 2-6 are presented results of measurements and calculations.

Table 2. Electrical parameters of tested solar cell with coating the investigated samples of solar glasses taken from nanostructurized side and from smooth side.

	Sample, No ; analysed side	I_{sc} [A]	V_{oc} [V]	P_{mp} [W]	FF [%]
	ref. cell	8,945	0,643	4,55	79,06
1:	nanostruct.	8,254	0,642	4,19	79,17
	smooth	8,253	0,642	4,20	79,16
2:	nanostruct.	8,238	0,642	4,187	79,15
	smooth	8,262	0,642	4,20	79,13
3:	nanostruct.	8,263	0,642	4,20	79,14
	smooth	8,266	0,642	4,20	79,10
4:	nanostruct.	8,276	0,642	4,20	79,18
	smooth	8,277	0,642	4,21	79,16
5:	nanostruct.	8,275	0,642	4,21	79,24
	smooth	8,274	0,642	4,21	79,21
6:	smooth/smooth	8,476	0,642	4,21	79,10

Table 3. Relative displacement voltage in the current – voltage curves ΔUoc. Electrical parameters of tested solar cell with coating the investigated samples of solar glasses taken from nanostructurized side and from smooth side.

Sample	I [A]	Uoc_nanost. [V]	Uoc_smooth [V]	ΔUoc
No. 1	0,5	0,63899	0,63933	-0,00034
	2	0,62997	0,63071	-0,00101
	4	0,61468	0,61593	-0,00125
No. 2	0,5	0,63909	0,63806	0,00103
	2	0,62940	0,62800	0,0014
	4	0,61480	0,61327	0,00153
No. 3	0,5	0,63905	0,63870	0,00035
	2	0,63010	0,62941	0,00069
	4	0,61484	0,61410	0,00074
No. 4	0,5	0,63947	0,63798	0,00149
	2	0,63011	0,62837	0,00174
	4	0,61534	0,61309	0,00225
No. 5	0,5	0,63914	0,63914	0
	2	0,62914	0,62954	-0,00040
	4	0,61548	0,61464	0,00084

Table 4. Transmittivity of investigated solar glasses samples for measurements taken from smooth side of sample.

Sample, No ; smooth side	$T_I = \frac{I_{SCR\,(1,2,3..)}}{I_{SCR\,(ref)}}$ [%]	$T_P = \frac{P_{MPR\,(1,2,3..)}}{P_{MPR\,(ref)}}$ [%]
ref. cell:	100,00	100,00
1:	96,50 ± 0,02	96,52 ± 0,02
2:	96,57 ± 0,03	96,55 ± 0,03
3:	96,62 ± 0,05	96,57 ± 0,05
4:	96,66 ± 0,01	96,65 ± 0,01
5:	96,70 ± 0,01	96,69 ± 0,01
6:	97,82 ± 0,93	96,90 ± 0,93

Table 5. Transmittivity of investigated solar glasses samples for measurements taken from nanostructurized side of sample.

nanost. side	$\varrho = \frac{S_{incl.}}{S_{sample}}$, [%]	$T_I = \frac{I_{SCR\,(1,2,3..)}}{I_{SCR\,(ref)}}$ [%]	$T_P = \frac{P_{MPR\,(1,2,3..)}}{P_{MPR\,(ref)}}$ [%]
ref. cell	-	100,00	100,00
1:	39	96,51 ± 0,09	96,42 ± 0,09
2:	73	96,45 ± 0,04	96,42 ± 0,04
3:	63	96,60 ± 0,03	96,57 ± 0,03
4:	28	96,65 ± 0,10	96,56 ± 0,10
5:	62	96,70 ± 0,02	96,69 ± 0,02
6:	0	97,82 ± 0,93	96,90 ± 0,93

Table 6. Absorption coefficient and calculated from formula (10) reflectivity coefficient R.

	Sample No ; analysed side	AI_{SCR} [%]	AP_{MPR} [%]	RI_{SCR} [%]	RP_{MPR} [%]
	ref. cell	0	0	0	0
1:	nanostruct.	0,93	0,95	2,56	2,63
	smooth	0,93	0,93	2,57	2,55
2:	nanostruct.	0,95	0,95	2,60	2,63
	smooth	0,91	0,92	2,52	2,53
3:	nanostruct.	0,90	0,91	2,50	2,52
	smooth	0,90	0,91	2,48	2,52
4:	nanostruct.	0,89	0,91	2,46	2,53
	smooth	0,89	0,89	2,45	2,46
5:	nanostruct.	0,88	0,88	2,42	2,43
	smooth	0,88	0,88	2,42	2,43
6:	smooth/smooth	0,58	0,83	1,60	2,27

Positive offset values indicate that higher light transmittance shows a rough surface solar glass (with a reflective coating), sometimes we observed greater transmissivity of the smooth side of the sample (Table 3 and Table 4). From certificate of sample done on QS130CA the differencies in Uoc between nanostructutized and smooth side of glasses ΔUoc is not visible (Table 2), but if we describe such differencies for different currents, we can see visible changes (Table 3). The biggest difference in voltage values estimated for the sample No. 4, the smallest - for the sample No. 5 (Table 3). All samples (except on the sample No.1) have a higher transmissivity for the roughened side of the glass (Table 5). Sample No.1 is the only one that has a superior transmission properties of the smooth surface of the sample than the nanostructuryzed surface.

We can noticed that the lower the density of the surface nanostructurized roughness (and absorption coefficient, Table 6), the higher the relative transmissivity of the glass (Table 5). The lower the height of the roughness (Table 1) the higher relative transmissivity of glass (Table 5).

The smaller the surface density and the smaller the roughness height the greater the measured voltage differences between smooth and rough surface of the glass. For all samples except sample 1, the anti-reflection layer exhibits a significantly better transmissivity (Table 5).

Contribution of reflection and absorption coefficient is 3.6% (Table 6), but the manufacturer's data shows that it is 0.7%.

The values of reflectance and transmittance should be treated as estimates, because they were measured without any integrating sphere.

* Corresponding author: pociask@ur.edu.pl

4 Conclusions

The relative changes in electrical solar cell parameters (short-circuit current, maximum power), it sets the relative transmissivity of solar glasses, are able to provide a test of quality of solar glasses.
From the information provided by D.A. Glass: the solar transmittance of glass is the best at range 60°-80° of angles of incidence of solar radiation. With help of QS130CA measurement were done with perpendicular angle of irradiation, and sample No. 6 for that reason (both smooth sides) is the best for such tests. It should be emphasized the role and importance of experimentally determined relative displacement voltage in the current – voltage curves. Transmissivity coefficient measured via presented way was acheved 96,7% for nanostructurized side of sample No.5. Our method shows that the best transmittancy has sample No.6 (without absorption coat), for perpendicular incidence of light to the test element.

References

1. www.daglass.pl/produkty/szklo-dekoracyjne
2. www.endeas.fi/download/QuickSun130CA.pdf
3. Kornelia Kalwas, METHODS FOR DETERMINATION OF TRANSMISSIVITY OF SELECTED MATERIALS USED AS PV ELEMENTS PROTECTIVE COVER, Faculty of Mathematics and Natural Sciences, University of Rzeszow, Rzeszow 2015, MSc thesis, under the supervisior of dr hab. prof. UR Malgorzata Pociask-Bialy

* Corresponding author: *pociask@ur.edu.pl*

Formation Dirac point and the topological surface states for HgCdTe-QW and mixed 3D HgCdTe TI

Michał Marchewka*

Faculty of Mathematics and Natural Sciences, Centre for Microelectronics and Nanotechnology, University of Rzeszów, Pigonia 1, 35-959 Rzeszów, Poland

Abstract. In this paper the results of numerical calculations based on the finite difference method (FDM) for the 2D and 3D TI with and without uniaxial tensile strain for mixed $Hg_{1-x}Cd_xTe$ structures are presented. The numerical calculations were made using the 8x8 model for x from 0 up to 0.155 and for the wide range for the thickness from a few nm for 2D up to 150 nm for 3D TI as well as for different mismatch of the lattice constant and different barrier potential in the case of the QW. For the investigated region of the Cd composition (x value) the negative energy gap ($E_g=\Gamma_8-\Gamma_6$) in the $Hg_{1-x}Cd_xTe$ is smaller than in the case of pure HgTe which, as it turns out, has a significant influence on the topological surface states (TSS) and the position of the Dirac point for QW as well as for 3D TI. The results show that the strained gap and the position of the Dirac point against the Γ_8 is a function of the x-Cd compounds in the case of the 3D TI as well as the critical width of the mixed $Hg_{1-x}Cd_xTe$ QW.

1 Introduction

In recent years, several different materials - topological insulators (TI) for which the topological surface states (TSS) were theoretically and experimentally confirmed have been identify. Starting from the 3D Bi-based materials e.g[1-5], through two dimensional (2D) systems based on HgCdTe/HgTe[6-8] up to strained 3D HgTe layers [9-11] all of these materials represent the new class of the quantum matter for which the metallic states exist on the surfaces. In all of these materials the conduction band is lower than the valence band which is caused by strong spin-orbital coupling. For all of these materials the TSS arise for different reasons [12,13]. The Dirac like dispersion is observed due to the potential of the 6.4 nm wide HgTe-QW in case of the CdTe/HgTe heterostructures. For 3D HgTe the uniaxial strain along (001) axis lifts the degeneracy of the Γ^8 band by breaking the cubic symmetry at the Γ-point and opens the insulating gap which makes such systems as 3D TI.

For 2D heterostructures as well as for the 3D TI based on the II-VI compounds like HgCdTe the eight band kp model is used for calculation the energy dispersion e.g [6,8,9,11,14-18]. This paper shows that the finite difference method (FDM) applied for 8×8 kp model [19,20] to solve numerical the differential Schrödinger equation gives an opportunity to analyse 2D and 3D structures which allows for creative design structures containing the so-called Dirac point and, what is important from the point of view of application possibilities - the shape of the Topological Protested Surface States (TPSS) [18].

For the two dimensional structures the experimental and the theoretical papers are devoted for the pure HgTe material as a material of the quantum wells. These structures are very good defined and the magneto optics and magetotransport properties have been investigated for a wide range of widths the HgTe-quantum well. For the mixed HgCdTe QW there is only a few papers which are dedicated such structures [21].

Similar case is for the 3D TI systems – the difference of the lattice constant between HgTe and the CdTe which is about 0.3% is enough to observe the TPSS and such structures are very popular e.g [9,18]. As for the 2D systems the 3D TI based on the HgTe are very popular and so far there is not so many papers devoted for 3D TI mixed HgCdTe [22].

There is still a few question e.g how to improve the 2D as well as the 3D structures to get the better properties of the electron transport through the interface/surface. For the 2D system the wider widths then 6.4 nm of the quantum wells seems to be better for such discussion. The detailed experiment were done for the 2D systems but for HgTe QW at different temperatures [23]. In the case of the 3D TI the situation is more complicated because the strained energy gap induced by the lattice mismatch is rather small (about 22meV) and only applied the external voltage it is possible to observe the dominating effects in electron transport thought the surface [18].

In this paper the results of numerical calculations for the unstrained and strained 2D and 3D (001) growth oriented mixed $Hg_{1-x}Cd_xTe$ structures are presented. The numerical calculations were made using the 8x8 kp model [19] for x from 0 up to 0.155 and for

* Corresponding author: marmi@ur.edu.pl

the different thickness (for QW and 3D TI-from 75 to 150 nm) as well as for a different mismatch of the lattice constant. For this region of the x-Cd composition the Γ^6 band approaching the degenerated $\Gamma_{hh}{}^8$ and $\Gamma_{lh}{}^8$ bands up to the values of the Eg = 0 (for x= 0.17 (at 4.2 K)) and for the 4um wide fully relaxed 3D crystal the Dirac like dispersion of the Γ^6 and $\Gamma_{lh}{}^8$ is expected [24]. Such structure with the unaxial strain for the investigated width region (up to 150 nm) is still strained [25]. The results for the dispersion relation for the typical 6.4 nm wide HgTe-QW as well as for the strained 75 nm wide HgTe structures obtained by the presented method are in a good the agreement with the results reported earlier [6,11,18].

As is presented below for the $Hg_{1-x}Cd_xTe$ unstrained quantum wells the critical widths is different for different x-Cd compounds as wells as in the well region as in the barrier materials. For the strained and wider than 6.4 nm 2D structures the electronic structures under the proper strain and x-Cd compounds looks like for the 3D TI based on mixed HgCdTe.

In the case of the 3D TI the numerical results show that the Dirac point can be found in the gap induced by strain in $Hg_{1-x}Cd_xTe$ structures with different x-Cd composition (for which the band structures are still inverted) and wider than in the earlier studies by other authors. The analysis of the dispersion relation for the 3D strained $Hg_{1-x}Cd_xTe$ layers is supplemented by the calculation of the dispersion relation for HgTe with a bigger unaxial strain than the one which usually comes from the different lattice constant between HgTe and CdTe. We demonstrate that the Dirac cone in 3D 75 nm wide HgTe cannot be found inside the indirect strained gap, not even for the bigger strain up to 2.3%. The combination of the different than zero x-Cd composition together with the bigger than 0.3% mismatch of the lattice constant between the substrate and the strained layers allow to get the dispersion relation of the TSS for 75 nm wide layers similar to those ones obtained for CdTe/HgTe QW with the most welcome position of the Dirac point inside the indirect isolating gap - a bit above the top of the $\Gamma_{hh}{}^8$ band.

2 Theory

2.1. Eight band kp model

The $Hg_{1-x}Cd_xTe$/HgTe QW are the heterostructures of type III[26] which depending on the width of the QW have normal or an inverted subband structures. For such materials the 8×8 kp model[19] (including spin splitting) allow, with successes, interpret a lot of experimental data. Because Schröodinger equation with the energy dependent effective mass with nonlinear eigenvalue problem cannot be solved directly, a few numerical methods (transfer matrix method[27-29], the shooting method[30], finite element method (FEM)[31] or FDM[32-35] have been proposed to solve the non-parabolic Schrödinger equation.

Numerical results presented in this paper were obtained using the 8x8 kp [19,20] model defined for (001) growth oriented structures. Originally, this model was applied to the quantum wells based on II-VI zinc-blende structures. In the presented kp model an eight band description of the band structure including all second-order terms representing the remote-band contributions with the J= +/- 3/2 - heavy hole and light hole bands – J=+/- 1/2 along with the J=+/- 1/2 conduction band states and J=+/-1/2 spin-orbit split off states are considered.

The Schröodinger equation for 8×8 Hamiltonian for the two dimensional vector of envelope functions, can be simply written as:

$$H\Psi(r) = E\Psi(r) \qquad (1)$$

where for such system the carrier wave-function can be written as:

$$\Psi(r) = \sum_n F_n(r) u_n(r). \qquad (2)$$

The $F_n(r)$ are the envelope function and the sum index of n varies from 1 to 8. The $u_n(r)$ are Bloch functions which are the same in the barrier and the well layers. F_n can be represented as

$$F_n = \exp\left[i\left(k_x x + k_y y\right)\right] f_n(z) \qquad (3)$$

where k_x and k_y are the wave vector components in the plane of the QW. Energy levels and envelope-function coefficients can be found from the simultaneous solution of the system of coupled differential equations. The Hamiltonian in such system takes the following form:

$$H_0 = \begin{bmatrix} T & 0 & -\frac{1}{\sqrt{2}}k_+P & \sqrt{\frac{2}{3}}k_zP & \frac{1}{\sqrt{6}}k_-P & 0 & -\frac{1}{\sqrt{3}}k_zP & -\frac{1}{\sqrt{3}}k_-P \\ 0 & T & 0 & -\frac{1}{\sqrt{6}}k_+P & \sqrt{\frac{2}{3}}k_zP & \frac{1}{\sqrt{2}}k_-P & -\frac{1}{\sqrt{3}}k_+P & \frac{1}{\sqrt{3}}k_zP \\ -\frac{1}{\sqrt{2}}k_-P & 0 & U+V & -\bar{S}_- & R & 0 & \frac{1}{\sqrt{2}}\bar{S}_- & -\sqrt{2}R \\ \sqrt{\frac{2}{3}}k_zP & -\frac{1}{\sqrt{6}}k_-P & -\bar{S}_-^{\dagger} & U-V & C & R & \sqrt{2}V & -\sqrt{\frac{3}{2}}\tilde{S}_- \\ \frac{1}{\sqrt{6}}k_+P & \sqrt{\frac{2}{3}}k_zP & R^{\dagger} & C^{\dagger} & U-V & \bar{S}_+^{\dagger} & -\sqrt{\frac{3}{2}}\tilde{S}_+ & -\sqrt{2}V \\ 0 & -\frac{1}{\sqrt{2}}k_+P & 0 & R^{\dagger} & \bar{S}_+ & U+V & \sqrt{2}R & \frac{1}{\sqrt{2}}\bar{S}_+ \\ -\frac{1}{\sqrt{3}}k_zP & -\frac{1}{\sqrt{3}}k_-P & \frac{1}{\sqrt{2}}\bar{S}_-^{\dagger} & \sqrt{2}V & -\sqrt{\frac{3}{2}}\tilde{S}_+ & \sqrt{2}R & U-\Delta & C \\ -\frac{1}{\sqrt{3}}k_+P & \frac{1}{\sqrt{3}}k_zP & -\sqrt{2}R^{\dagger} & -\sqrt{\frac{3}{2}}\tilde{S}_-^{\dagger} & -\sqrt{2}V & \frac{1}{\sqrt{2}}\bar{S}_+^{\dagger} & C^{\dagger} & U+\Delta \end{bmatrix} \qquad (4)$$

where

$$k_{||}^2 = k_x^2 + k_y^2;\ k_\pm = k_x \pm k_y;\ k_z = -\frac{i\partial}{\partial z};$$

$$T = E_c(z) + \frac{\hbar^2}{2m_0}\left[(2F+1)k_{||}^2 + k_z(2F+1)k_z\right]$$

$$U = E_v(z) - \frac{\hbar^2}{2m_0}\left(\gamma_1 k_{||}^2 + k_z\gamma_1 k_z\right)$$

$$V = -\frac{\hbar^2}{2m_0}\left(\gamma_2 k_{||}^2 + 2k_z\gamma_2 k_z\right)$$

$$R = -\frac{\hbar^2}{2m_0}\sqrt{3}(\mu k_+^2 - \bar{\gamma}k_-^2)$$

$$\bar{S}_\pm = -\frac{\hbar^2}{2m_0}\sqrt{3}k_\pm(\{\gamma_3, k_z\} + [\kappa, k_z])$$

$$\tilde{S}_\pm = -\frac{\hbar^2}{2m_0}\sqrt{3}k_\pm\left(\{\gamma_3, k_z\} - \frac{1}{3}[\kappa, k_z]\right)$$

$$C = \frac{\hbar^2}{2m_0}k_-[\kappa, k_z] \qquad (5)$$

 DOI: 10.1051/epjconf/201713304001

[A,B]=AB-BA is the usual commutator and {A,B}=AB+BA is the anticommutator for the operators A and B; P is the Kane momentum matrix element;
$E_c(z)$ and $E_v(z)$ are the conduction and valence band edges, respectively; Δ is the spin-orbit splitting energy; γ_1, γ_2, γ_3, κ and F describe the coupling to the remote bands and the μ and $\bar{\gamma}$ are parameters according to $\mu=(\gamma_3-\gamma_2)/2$ and $\bar{\gamma}=(\gamma_3+\gamma_2)/2$. The band structure parameters for HgTe and CdTe used in calculations are listed in Table I. The dependence of the band gap for

$Hg_{1-x}Cd_xTe$ as a function of the temperature and composition x is determined from the empirical expression according to C. R. Becker et al. [27]. The valence band offset between HgTe and CdTe is taken to be equal to 570 ± 26 meV at T=5 K [27]. In the calculation is used 570 meV. The strain tensor components for an arbitrary growth direction can be determined by using the model of De Caro et al. [37]. The effects of the strain tensor can be incorporated in the Kane model through the Bir-Pikus Hamiltonian[38], which can be easily obtained from the Kane Hamiltonian with the substitution $k_i k_j \rightarrow \varepsilon_{ij}$. The shape of the Bir-Pikus Hamiltonian used in the calculations has been described in detail by E. G. Novik[20]. The strain tensor ε_{ij} for (001) growth direction has three values different than 0: $\epsilon_{11}, \epsilon_{22}$ and ϵ_{33} where $\epsilon_{11} = \epsilon_{22} = \epsilon_{33} = \epsilon_{||}$, and $\epsilon_{33} = -2\varepsilon C_{12}/C_{11}$. The lattice mismatch is defined as $\epsilon = (a_{CdTe} - a_{HgTe})/(a_{HgTe})$, and for the HgTe it is about 0.3%. The band gap between Γ_{hh}^8 and Γ_{lh}^8 is about 22 meV at k=0 due to the value of ϵ for (001) for the strained HgTe structure. The lattice constant of $Hg_{1-x}Cd_xTe$ as a function of x was taken as[39]: $a = 6.4614 + 0.008x + 0.01168x^2 - 0.0057x^3$ (in terms of angstroms). The mismatch of the lattice constant is the main element which affects a size of the gap for the $Hg_{1-x}Cd_xTe$ for which the Cd composition is between the studied region (which is about is about 20 meV see Fig. 1). In Fig. 1 the strained energy gap as a function of the x-Cd compounds are presented in two cases: green curves shows the approximation of the Eg(x-Cd) as a function of a(x) only while the red curves presents the Eg(x-Cd) when the Eg is a function of lattice constant as wells as the confection of the deformation potential and the stiffness constants. In the investigated region of x-Cd compounds (0 - 0.155 see inset) the Eg can be fitted using the approximation plotted by the green curve.

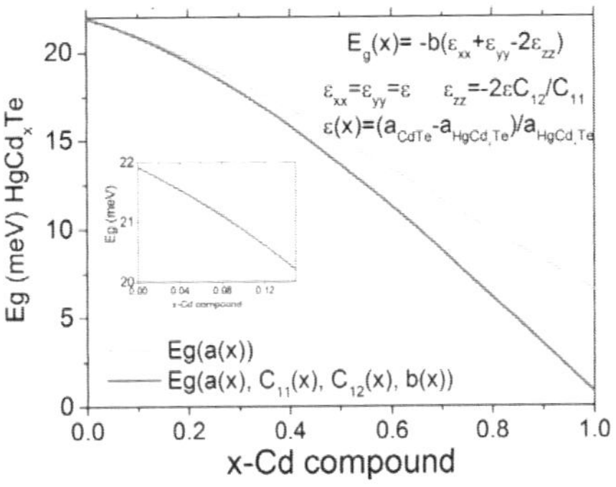

Fig. 1. The strained gap (between Γ_{hh}^8 and Γ_{lh}^8) as a function of: the lattice constant (green curve) only and the confection of the deformation potential and the stiffness constants (red curve) for the different x-Cd compounds

The results presented below were received for the band structure parameters collected in Table 1. Besides the coupling to the remote bands parameters (see above) we have: a, b, d -the Bir-Pikus deformation potential[38]; C, C_{11}, C_{22}, C_{44} - the stiffness constants. All presented results were calculated for 4.2 K and for the $k_{||} = (k_x, 0)$.

Table 1. The parameters used in calculations

	E_g	E_v	Δ	$E_p=2m_0P^2/\hbar^2$	F	γ_1	γ_2	γ_3	κ
HgTe	-0.303 eV	0	1.08 eV	18.8 eV	0	4.1	0.5	1.3	-0.4
CdTe	1.606 eV	-570 meV	0.91 eV	18.8 eV	-0.09	1.47	-0.28	0.03	-1.31

3 Results

3.1. Quantum wells

All presented results for quantum wells are obtained for the case when both barriers are about 100 nm which in practice means infinite potential. To get the better agreement with the real structures of the QW three different region were taken into account in the numerical calculation (see fig. 2): the CdTe layer (70 nm) and the buffer layer $Hg_{0.3}Cd_{0.7}Te$ (30 nm) which are placed for both sides of the different width of the HgTe-QW.

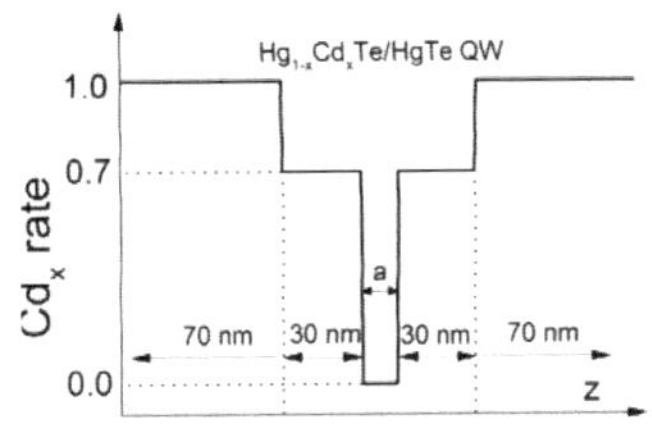

Fig. 2. The shape of potential of the investigated QW structures.

The distance between nearest two points Δz were defined to be equal half of the lattice constant of a unit cell. The dependencies of all parameters, except the band-gap, on the content of the solid solution $Hg_{1-x}Cd_xTe$ are assumed to be linear in x. To verify the validity of our proposed method and find the application range of this approach, the dispersion relation energy spectrum for different width of the QW were calculated.
The dispersion relation for the potential profile presented in Fig. 3 were calculated for two direction of k: $k_{||} = [10]$and $k_{||} = [11]$ HgTe QW for different values of the width of the QW and the results are presented in Fig. 4.

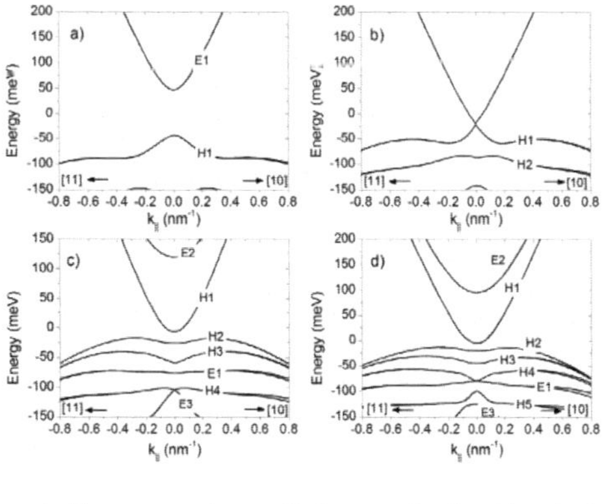

Fig. 3. Energy spectrum of size-quantized subbands in a symmetric CdTe/ $Hg_{0.3}Cd_{0.7}Te$/HgTe QW for a) 4nm, b) 6nm, c) 12nm, d) 14nm – wide QW calculated by FDM method applying for 8x8 kp Kane Hamiltonian for: $k_{||} = [10]$and $k_{||} = [11]$.

It can be clearly seen that our results are in good agreement with this presented in the literature by others authors for the heterostructures with normal band structures e.g [40] as well as with inverted band structures e.g [41]. The case of the pure HgTe is very well known but shows that the simply FDM method is very good tools for this purpose.
The different then (001) grow oriented structures can be calculated using the Hamiltoanin for zinc-blende structures presented in Ref. [19]. In the case of the 6nm HgTe QW (Fig. 4a) the difference between these two oriented structures in visible only for Γ_{lh} subband and only for k>0.1nm^{-1}. The bigger difference is observed for strained QW (Fig. 4b). Such wide structures is strongly investigated [42]. In this case the strained undirected energy gap is very small (<1meV) for 14 nm strained HgTe QW. It is worth to mention that the self-consistent potential is not included in presented results. Such potential can play an important role for the case when the undirect gap is so small like in the presented case.

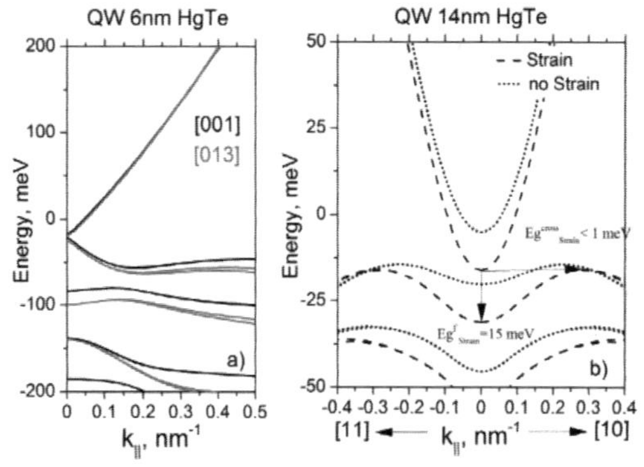

Fig. 4 . The dispersion relation for: a) 6nm wide HgTe QW for two different grow oriented structures- [001] and [013]; b) for strained and unstrained 14 nm wide HgTe QW.

The question is if it is possible to obtain a strained QW with Dirac point, and how does the dispersion relation look like in such cases. The answer is presented on Fig. 5b, where the $(E(k_{||})$ for a different HgTe QWs width was calculated for different mismatch levels of lattice constants so that each time a Dirac cone can be achieved for each QWs. The wider QW needs bigger lattice mismatch (see Fig. 5a), the bigger lattice mismatch caused that the light hole sub band is changes so for such structures with small concentration it is possible to observe the hole behaviour in for which the dispersion have Dirac like shape.

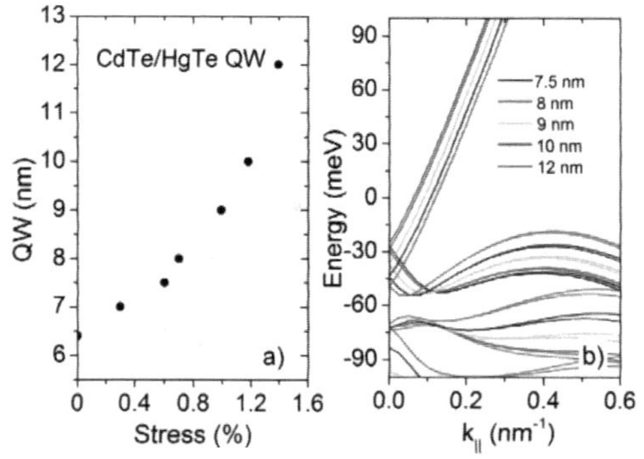

Fig. 5. a) the QW width as a function of the strain (lattice mismatch) for structures with Dirac point: b) The $E(k_{||})$ for a different HgTe QWs width was calculated for different mismatch levels of lattice constants so that each time a Dirac cone can be achieved for each QWs. The parameters of the lattice mismatch for the following QWs are equal: 0.32% for 7.5 nm wide QW, 0.40% for 8 nm wide QW, 0.70% for 9 nm wide QW, 0.98% for 10 nm wide QWand 1.20% for 12 wide nm QW.

The another possibility is to get the two dimensional Dirac like dispersion for mixed HgCdTe structures. For the case when the barrier area are build by HgCdTe it is possible to obtain the Dirac point but for the QW which are thinner then critical value 6.4nm for pure HgTe (see Fig. 6a). On the other hand, for the mixed material for QW the Dirac like dispersion for unstrained structures can be received for wider QW structures – about 10nm for $Hg_{0.95}Cd_{0.05}Te$ QW and 17nm for $Hg_{0.9}Cd_{0.1}Te$ QW.

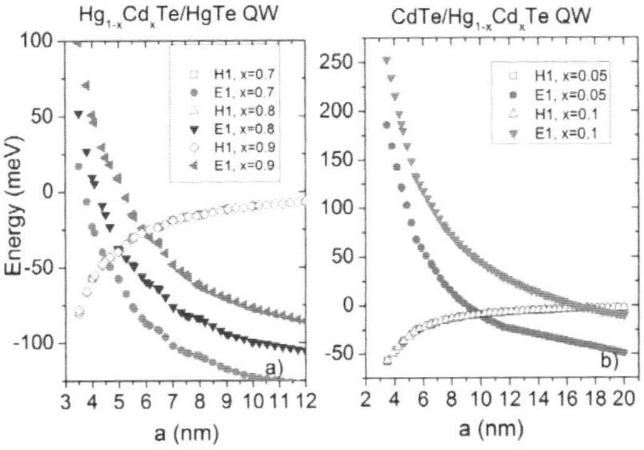

Fig. 6. Energy of the electron like subband (E1) and like – hole (H1) as a function of the widths the QW for: a) the different x-Cd compounds in the barrier material in HgCdTe/HgTe-QW and b) different x-cd compounds in the QW material in CdTe-HgCdTe QW.

3.2 3D strained systems

Dispersion relation calculated within an eight-band kp model at eight different x-Cd composition including strain for $Hg_{1-x}Cd_xTe$ layers 150 nm wide are presented in Fig. 7. In case of the HgTe (Fig. 7a)) the dispersion relation looks the same as in the other publications for the 75 nm wide structures [7,18]. It should be noted that because of hybridization with the valence-band states, the position of the Dirac point for HgTe is not directly obvious from the dispersion plot in the presented range - it is located at about -81 meV. In the presented cases the calculations were made for symmetric structures with two identical interfaces: $CdTe/Hg_{1-x}Cd_xTe$ and $Hg_{1-x}Cd_xTe/CdTe$ so there is no inversion symmetry break and the Dirac point does not split as it happens in case of two different interfaces.

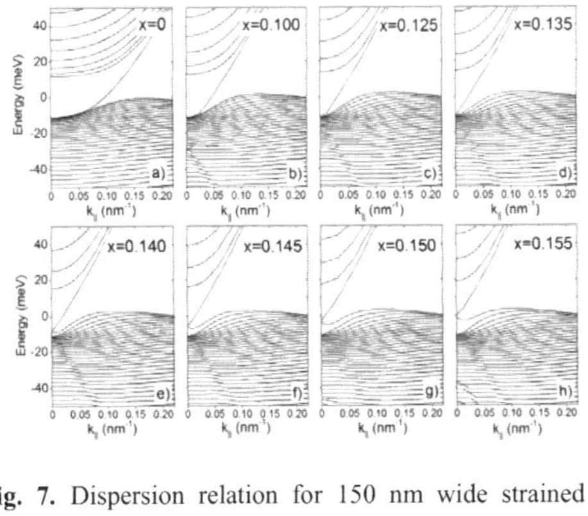

Fig. 7. Dispersion relation for 150 nm wide strained 3D $Hg_{1-x}Cd_xTe$ structures for x from 0 up to 0.155.

The position of the Dirac point goes up together with the increase of the x-Cd composition. For x=0.100 is located around -29 meV (Fig. 7b)), for the higher values of x Dirac point is located around the top of the valence band (for x=0.125 between -20 and -15 meV - Fig. 7c), and about -11 meV for x=0.135 - Fig. 7d). For the higher value of the x the Dirac point is being created inside the energy gap induced by the strain (Figs. 7e)-h)). For these values of the Cd composition the $Hg_{1-x}Cd_xTe$ is a true topological insulator. The information about the location of the Dirac cone in each cases can be inferred from the distribution of the charge carriers. As mentioned above, two identical interfaces are considered so there is one Dirac point as it can be seen in Figs. 7b) and 7c).

It can be clearly seen in Fig. 7 that for x>0.130 the energy of the surface states can be found in the fullest extent of the energy in the strained gap. For the x=0.155 the $\sum_{i=1}^{8}\Psi_i^2(z)$ were calculated (see Fig. 8) along $k_{||}$for the energy dispersion region inside the strained gap which corresponds to $k_{||}$from $k_{||}$ =0 up to $k_{||} = 0.06nm^{-1}$. The detailed description of the charge distribution in mixed $Hg_{1-x}Cd_xTe$ 3D TI is presented in Ref. [43].

The question is if it possible to obtain the Dirac point at the energy gap for 3D HgTe TI. Our previous calculation was done for 150 nm wide $Hg_{1-x}Cd_xTe$ structures with x=0.155 and 0.3% of the mismatch of the lattice constant due to the fact that for this type of structure, the Dirac point is in the middle of the strained gap - Fig. 8h.

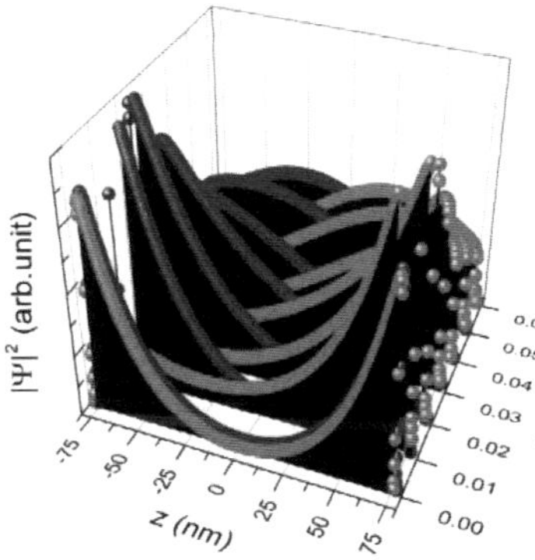

Fig. 8. $\sum_{i=1}^{8}\Psi_i^2(z)$ along $k_{||}$ for the energy dispersion region inside the strained gap which corresponds to $k_{||}$ from $k_{||}$ =0 up to $k_{||} = 0.06nm^{-1}$for two energy states for the upper part of the Dirac cone (blue and red point respectively) from $E(k_{||} = 0) = 0$ meV up to $E(k_{||} = 0.06nm^{-1}) = 20$ meV for 150nm wide strained 3D $Hg_{0.845}Cd_{0.155}Te$.

The detailed analysis show that for the pure 3D HgTe 75nm wide TI it is possible to obtain the Dirac point inside of the strained gap only if the mismatch of the lattice constancy is equal at least 5% [43]. In the opposite to this the Dirac point inside of the strained gap can be found for mixed HgCdTe but the width of the structures has to be about 150 nm – for the narrower structures the gap between two Dirac cones appears and the charge distribution is typical like for the bulk.

DOI: 10.1051/epjconf/201713304001

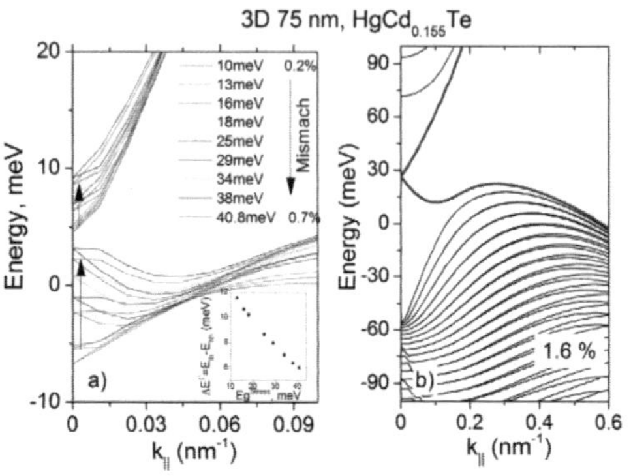

Fig. 9. a) $E(k_{||})$ for different lattice mismatch for 3D 75nm wide HgCd0.155Te (inset the difference between upper and lower Dirac cones as a function of the gap induced by strain is presented); b) the 75nm $HgCd_{0.155}Te$ wide 3D TI with lattice mismatch equal 1.6% - there is no gap between Dirac cones.

On Fig. 9 the analysis of the strained 75 nm wide $HgCd_{0.155}Te$ are presented. To get the graphene like dispersion in such structures it is necessary to applied the strain which can be realized by 1.6% lattice mismatch between substrate and 3D TI material (Fig. 9b). In other cases (smaller strain) (Fig. 9a) the gap between Dirac cones caused that the charge distribution is typical like for the bulk[43].

4 Summary

In the presented papers the results of the calculation the dispersion relation for mixed HgCdTe 2D and 3D structures are presented. The uniaxial tensile strain and the different x-Cd composition cause that obtained Dirac like dispersion in those cases can be achieve for different mixed heterostructures. For the 2D mixed $Hg_{1-x}Cd_xTe$ for different widths which is caused by the different level of the x-Cd compounds in the quantum well material as wells as in the barrier material the Dirac point can be found also. Our calculations show that it is possible to obtain the dispersion relation of the TSS for 3D 75 nm wide strained layers which looks similar to that one obtained for the strained HgCdTe/HgTe QW with the range width from 8 nm up to 9 nm. It happens if two conditions are fulfilled: the x-Cd composition should be equal about 0.155, and the level of the mismatch of the lattice constant should be about 1.6% (Fig. 7c). Then the Dirac point at the Γ-point is above the top of the Γ^8_{hh} band. The consequences of such shape of the dispersion relation of the TSS for the investigated 2D and 3D strained system are highly desirable. In the case of the 3D system the energy distance between Dirac point and the bottom of the Γ^8_{lh} is much greater than in case of the typical 3D HgTe TI and is over 40 meV. The shape of the lower part of the Dirac cone and its relative position to the Γ^8_{hh} states gives an opportunity for the experimental investigation of this TSS part as it is also for the strained 2D system. As shown by the other authors the theoretical and experimental investigation of both upper and lower part of the Dirac cones in case of the HgTe QW provides a lot of interesting results [44-48]. The 2D and 3D strained structures with the TSS dispersion relation similar to that obtained for HgTe QW should be a very interesting experimental object for further verification of all these effects.

This work was partly supported by the EU Foundation by the TARI-contract PRI-CT-1999-00088. We acknowledge support from the authorities of the Podkarpackie Voiwodership (Marshals O_ce of the Podkarpackie Voivodship of Poland), contract WNDPPK. 01.03.00-18-053/12

References

1. L. Fu, C. L. Mele, Phys. Rev. Lett. **94** (2007)
2. H. Zhang, C. X. Liu, X. L. Qi, X. Dai, Z. Fang and S. C. Zhang, Nature Physics **5**, 438-442 (2009)
3. D. Hsieh, D. Qian, L. Wray, Y. Xia, Y. S. Hor, R. J. Cava, M. Z. Hasan, Nature. **452**, 970-974. (2008)
4. Y. Xia, D. Qian, D. Hsieh, L. Wray, A. Pal, H. Lin, A. Bansil, D. Grauer, Y. S. Hor, R. J. Cava, M. Z. Hasan, Nature Physics. **5**, 398-402 (2009)
5. D. Hsieh, Y. Xia, D. Qian, L. Wray, F. Meier, J. H. Dil, J. Osterwalder, L. Patthey, A. V. Fedorov, H. Lin, A. Bansil, D. Grauer, Y. S. Hor, R. J. Cava, and M. Z. Hasan, Phys. Rev. Lett. **103,** 146401 (2009)
6. B. A. Bernevig, T. L. Hughes, S. C. Zhang, Science **314,** 1757 (2006)
7. M. Konig, S. Wiedmann, C. Brüne, A. Roth, H. Buhmann, L. W. Molenkamp, X. L. Qi, S. C. Zhang. Science **318,** 5851 (2007)
8. M. Konig, H. Buhmann, L. W. Molenkamp, T. L. Hughes, C.X. Liu, X. L. Qi, S. C. Zhang. J. Phys. Soc. Jap. **77,** 031007 (2008)
9. C. Brüne, C. X. Liu, E. G. Novik, E. M. Hankiewicz, H. Buhmann, Y. L. Chen, X. L. Qi, Z. X. Shen, S. C. Zhang, and L.W. Molenkamp. Phys. Rev. Lett. **106**, 126803 (2011)
10. D. A. Kozlov, Z. D. Kvon, E. B. Olshanetsky, N. N. Mikhailov, S. A. Dvoretsky, and D. Weiss. Phys. Rev. Lett. **112**, 196801 (2014)
11. K. M. Dantscher, D. A. Kozlov, P. Olbrich, C. Zoth, P. Faltermeier, M. Lindner, G. V. Budkin, S. A. Tarasenko, V. V. Belkov, Z. D. Kvon, N. N. Mikhailov, S. A. Dvoretsky, D. Weiss, B. Jenichen, and S. D. Ganichev. Phys. Rev. B. **92,** 165314 (2015)

12. M. Z. Hasan and C. L. Kane. Rev. Mod. Phys. **82**, 3045 (2010)
13. X. L. Qi and S. C. Zhang. Rev. Mod. Phys. **83**, 1057 (2011)
14. X. Dai, T. L. Hughes, X. L. Qi, Z. Fang, and S. C. Zhang. Phys. Rev. B. **77**, 125319 (2008)
15. C. X. Liu, X. L. Qi, H. J. Zhang, X. Dai, Z. Fang, and S. C. Zhang. Phys. Rev. B. **82**, 045122 (2010)
16. D. G. Rothe, R. W. Reinthaler, C. X. Liu, L. W. Molenkamp, S. C. Zhang and E. M. Hankiewicz. New Journal of Physics **12**, 065012 (2010)
17. R. L. Chu, W. Y. Shan, J. Lu, and S. Q. Shen. Phys. Rev. B. **83**, 075110 (2011)
18. C. Brune, C. Thienel, M. Stuiber, J. Bottcher, H. Buhmann, E. G. Novik, C. X. Liu, E. M. Hankiewicz, L. W. Molenkamp. Phys. Rev. X. **4**, 041045 (2014)
19. A. Pfeuffer-Jeschke, *Bandstruktur und landau-Niveaus quecksilberhaltinger II-VI-Heterostrukturen*, Universitt Wurzburg, Germany: Physikalisches Institut; (2000)
20. E. G. Novik, A. Pfeuffer-Jeschke, T. Jungwirth, V. Latussek, C. R. Becker, G. Landwehr, H. Phys. Rev. B. **72**, 035321 (2005)
21. S. A. Dvoretsky, D. G. Ikusov, Z. D. Kvon, N. N. Mikhailov, V. G. Remesnik, R. N. Smirnov, Yu. G. Sidorov, V. A. Shvets, *Semiconductor Physics, Quantum Electronics & Optoelectronics* **10**, 47-53 (2007)
22. G. Tomaka, J. Grendysa, P. Sliz, C. R. Becker, J. Polit, R. Wojnarowska, A. Stadler, E. M. Sheregii. Phys. Rev. B. **93**, 205419 (2016)
23. S. Wiedmann, A. Jost, C. Thienel, C. Brune, P. Leubner, H. Buhmann, L. W. Molenkamp, J. C. Maan, and U. Zeitler. Phys. Rev. B. **91**, 205311 (2015)
24. M. Orlita, D. M. Basko, M. S. Zholudev, F. Teppe, W. Knap, V. I. Gavrilenko, N. N. Mikhailov, S. A. Dvoretskii, P. Neugebauer, C. Faugeras, A. L. Barra, G. Martinez, M. Potemski, Nature Physics **10**, 233-238 (2014)
25. O. Crauste, Y. Ohtsubo, P. Ballet, P. A. Louis Delplace, D. Carpentier, C. Bouvier, T. Meunier, A. Taleb-Ibrahimi, L. Lévy. arXiv.:1307 (2008)
26. N. F. Johnson, H. Ehrenreich, P. M. Hui, P. M. Young. Phys. Rev. B. **41**, 3655 (1990)
27. L. R. Ram, K. H. Yoo, R. L. Aggarwal. Phys. Rev. B. **38**, 6151 (1988)
28. K. H. Yoo, L. R. Ram-Mohan, and D. F. Nelson. Phys. Rev. B. **39**, 12808 (1989)
29. D. C. Hutchings. Apply. Phys. Lett. **55**, 1082 (1989)
30. P. Harrison. *QuantumWells, Wires and Dots*. 2nd ed. Wiley, New York (2006)
31. K. Q. Le, *Microwave and Optical Technology Letters*. **51**, 1 (2009)
32. X. Ma, K. Li, Z. Zhang, H. Hu, Q. Wang, X. Wei, G. Song. Journal of Apply. Phys. **114**, 063101 (2013)
33. S. F. Tsay, J. C. Chiang, Z. M. Chau, I. Lo. Phyes. Rev. B. **56**, 13242 (1997)
34. J. D. Cooper, A. Valavanis, Z. Ikoni, P. Harrison, J. E. Cunningham. Apply. Phys. Lett. **108**, 113119 (2010)
35. Y. Jiang, X. Ma, Y. Xu, G. Song. Journal of Apply. Phys. **116**, 173702 (2014)
36. C. R. Becker, V. Latussek, A. Pfeuffer-Jeschke, G. Landwehr, L. W. Molenkamp. Phys. Rev. B. **62**, 10353 (2000)
37. L. D. Caro, L. Tapfer. Phys Rev. B. **51**, 4374 (1995)
38. G. L. Bir, G. E. Pikus. *Symmetry and Strain-Induced Effects in Semiconductors*. New York: Wiley; (1974)
39. A. Rogalski. Infrared Photon Detectors. Bellingham: SPIE Optical Engineering Press; (1995)
40. M. Zholudev, F. Teppe, M. Orlita, C. Consejo, J. Torres, N. Dyakonova, M. Czapkiewicz, J. Wróbel, G. Grabecki, N. Mikhailov, S. Dvoretskii, A. Ikonnikov, K. Spirin, V. Aleshkin, V. Gavrilenko, and W. Knap. Phys. Rev. B. **86**, 205420 (2012)
41. G. M. Minkov, A. V. Germanenko, O. E. Rut, A. A. Sherstobitov, S. A. Dvoretski, and N. N. Mikhailov. Phys. Rev. B. **93**, 155306 (2013)
42. E. B. Olshanetsky, Z. D. Kvon, G. M. Gusev, A. D. Levin, O. E. Raichev, N. N. Mikhailov, and S. A. Dvoretsky. Phys. Rev. Lett. **114**, 126802 (2015)
43. M. Marchewka, Physica E. **84**, 407-414 (2016)
44. M. V. Durnev and S. A.Tarasenko. Phys. Rev. B. **93**, 075434 (2016)
45. G. M. Minkov, A. V. Germanenko, O. E. Rut, A. A. Sherstobitov, M. O. Nestoklon, S. A. Dvoretski, and N. N. Mikhailov. Phys. Rev. B. **93**, 155304 (2016)
46. S. A. Tarasenko, M. V. Durnev, M. O. Nestoklon, E. L. Ivchenko, J.W. Luo, and A. Zunger. Phys. Rev. B. **91**, 081302(R) (2015)
47. O. E. Raichev, G. M. Gusev, E. B. Olshanetsky, Z. D. Kvon, N. N. Mikhailov, S. A. Dvoretsky, and J. C. Portal. Phys. Rev. B. **86**, 155320 (2012)
48. M. Pang and X. G. Wu. Phys. Rev. B. **88**, 235309 (2013)

Calculation of thermodynamic functions of saturated solid solution of $AgIn_2Te_3I$ compound in the Ag–In–Te–I system

Mykola Moroz[1], *Sergiy* Prokhorenko[2,*], *Myroslava* Prokhorenko[3] and *Oleksandr* Reshetnyak[4]

[1]Department of Chemistry and Physics, National University of Water and Environmental Engineering, 33000 Rivne, Ukraine
[2]Center for Microelectronics and Nanotechnology. University of Rzeszow, 35-959 Rzeszów, Poland
[3]Department of Cartography and Geospatial Modeling, Lviv Polytechnic National University, 79000 Lviv, Ukraine
[4]Department of Physical and Colloid Chemistry, Ivan Franko National University of Lviv, 79000 Lviv, Ukraine

Abstract. Triangulation of Ag–In–Te–I system in the vicinity of $AgIn_2Te_3I$ compound was investigated by X-ray diffraction and differential thermal analysis methods. The spatial position of the phase region $AgIn_2Te_3I$–InTe–Ag_2Te–AgI regarding the figurative point of silver was used in order to write the equation of virtual potential-forming reaction. Potential-forming reaction was performed in electrochemical cell (ECC) of the type (–) C | Ag | $Ag_3GeS_3I(Br)$ glass | *D* | C (+) where C are inert (graphite) electrodes; Ag and *D* are the electrodes of the ECC; *D* represents the alloy of four-phase region; Ag_3GeS_3I glass is a membrane with purely ionic Ag^+ conductivity). Linear dependence of the EMF of cell on temperature in the range of 440–480 K was used to calculate the standard thermodynamic functions of saturated solid solution of $AgIn_2Te_3I$ compound in Ag–In–Te–I system.

1 Introduction

In recent years, there has been an increasing interest in solid ionic conductors in view of the need to develop novel devices such as micro- and nano supercapacitors, ion selective membranes, batteries, fuel cells, sensors, electrochromic displays, etc [1, 2]. The $AgIn_2X_3Y$ compounds (X – S, Se, Te; Y – Cl, Br, I) belong to the group of silver-containing chalcogenide crystalline $Ag_3SI(Br)$ and glassy $Ag_3GeS_3I(Br)$ superionic phases in which chalogenide anions, along with silver cations, possess the properties of non-stationary quasi-liquids [3].

This paper presents the results of the experimental determination of the boundaries of phase regions including the $AgIn_2Te_3I$ compound, and calculation of thermodynamic functions of saturated solid solution of the four-element compound.

2 Experimental

The InTe, In_2Te_3, Ag_2Te compounds were prepared by cooling of the melts of the mix of the elements of semiconductor purity. The Ag_3GeS_3I glass was obtained by quenching the melt of elements and AgI from 1200 K into ice water. Crystals of indium and silver tellurides (particle size ≤ 5 μm) together with powdered AgI were used for making the positive electrodes of electrochemical cell (ECC) and for alloys for differential thermal analysis (DTA) and X-ray diffraction (XRD). Well-mixed blends of calculated quantities of the compounds were placed into quartz ampoules and evacuated to the residual pressure ~ 1 Pa. Solid-phase synthesis was performed at 820 K for 20 days. The phase composition of the alloys was determined by DTA and XRD. XRD patterns were collected on a STOE STADI P diffractometer [4] equipped with a linear position–sensitive detector PSD, in a modified Guinier geometry (transmission mode, $CuK\alpha_1$ radiation, a bent Ge (111) monochromator, $2\theta/\omega$ scan mode). XRD arrays were processed using STOE $WinX^{POW}$ (version 2.21) and PowderCell (version 2.3) software suits.

Potential-forming process was performed in electrochemical cell of the type (–) C | Ag | Ag_3GeS_3I glass | *D* | C (+) where C are the inert graphite electrodes; Ag and *D* are the electrodes of ECC; *D* represents the alloy of four-phase region; and Ag_3GeS_3I glass is a membrane with purely ionic Ag^+ conductivity.

Powdered cell components were pressed ($p \sim 10^8$ Pa) into through holes with the diameter of 2 mm arranged in the PTFE matrix up to the density $\rho = (0.93 \pm 0.02)\rho_0$, where ρ_0 is the experimentally determined density of cast alloys. To eliminate the defects of plastic deformation under extrusion of alloys, we performed five-fold thermal cycling of ECC in the range of 400–480 K with heating and cooling rates of 2 K min^{-1}. The ECC was heated in a resistance furnace filled with a mixture of H_2 and Ar in a molar ratio of 1 : 9, $p = 1.2\times10^5$ Pa. The flow of gas at the rate of 2×10^{-3} m^3 h^{-1} had a direction from the positive to the negative electrode of the ECC. The temperature was maintained with an accuracy of ±0.5 K. The EMF values of the cells were measured using the voltmeter of a U7–9 electrometric amplifier with an input resistance of $>10^{12}$ Ω.

* Corresponding author: siergiej.prokhorenko@gmail.com

DOI: 10.1051/epjconf/201713304002

3 Results and discussion

According to Fig. 1, virtual chemical reaction in the ECC involving four-element phase is described by the equation

$$2Ag + AgIn_2Te_3I = AgI + 2InTe + Ag_2Te. \quad (1)$$

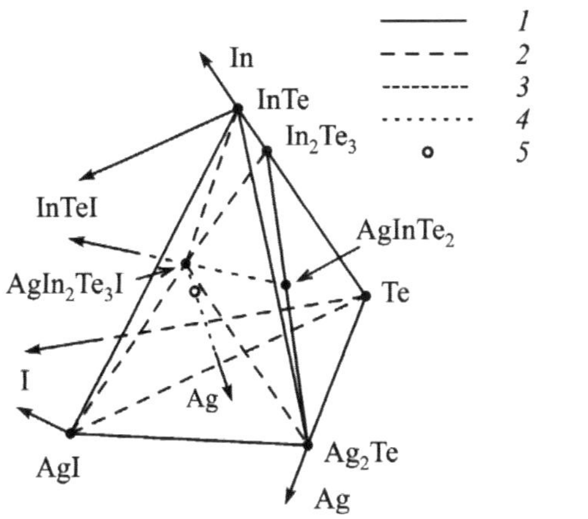

Fig. 1. Triangulation of the Ag–In–Te–I system in the vicinity of $AgIn_2Te_3I$ compounds: *1, 2* are the lines of two-phase equilibria; *3, 4* are the sections Ag–$AgIn_2Te_3I$ and $AgInTe_2$–InTeI; *5* is the figurative point of the alloy of the positive electrode of ECC.

The equation is written in the approximation of little difference between the position of figurative point of the compound and of its saturated solid solutions in Gibbs tetrahedron. Based on Eq. (1) the positive electrode of ECC was formed from the mixture of powdered silver and compounds AgI, In_2Te_3, in the molar ratio Ag : AgI : In_2Te_3 = 1 : 1 : 1. The linear dependence EMF (E) on temperature (T) for ECC with positive electrodes D in the range 440–480 K is approximated by the equation

$$E\,/\,\text{mV} = (164.67971 \pm 0.86451) + (0.16850 \pm 0.00186)T\,/\,\text{K}. \quad (2)$$

The deviation from the linearity above 480 K is likely due to the changes in the regions of the existence of the phases.

From the measured $E(T)$ relation, Gibbs energy was calculated using the basic thermodynamic equation

$$-\Delta_r G_T = n\,\text{F}\,E(T), \quad (3)$$

where $n_e = 2$ is a number of electrons taking part in the reaction (1); F = 96485.3 C mol^{-1} is Faraday's number.

According to (1)

$$\Delta_r G_{T,(1)} = 0.5\Delta_f G_{T,\beta\text{-}Ag_2Te} + \Delta_f G_{T,InTe} + 0.5\Delta_f G_{T,\beta\text{-}AgI} - 0.5\Delta_f G_{T,\beta\text{-}AgIn_2Te_3I} \quad (4)$$

where $\Delta_f G_T$ are the Gibbs energies of the formation of the appropriate phase from the elements at temperature T.

From Eq. (2) the Gibbs energy of the formation of $AgIn_2Te_3I$ is then:

$$\Delta_f G_{T,AgIn_2Te_3I} = \Delta_f G_{T,\beta\text{-}Ag_2Te} + 2\Delta_f G_{T,InTe} + \Delta_f G_{T,\beta\text{-}AgI} - 2\Delta_r G_{T,(1)} \quad (5)$$

Considering that $\Delta_f G_{T,\,Ag2Te}$=(−26.0129±1.858)–(42.8±9)×10^{-3}T [5], $\Delta_f G_{T,\,InTe}$= −71.965+1.634×10^{-3}T, $\Delta_f G_{T,\,\beta\text{-}AgI}$ = −63.214 – 14.19×10^{-3}T [6], and using Eqs. (2) and (3) we obtained

$$\Delta_f G_{T,\,AgIn2Te3I}/(\text{kJ mol}^{-1}) = (-201.4 \pm 1.9) - (21.2 \pm 9.5) \times 10^{-3}\,T/\text{K}. \quad (6)$$

Eq. (6) expresses the temperature dependence of the Gibbs energy of the formation of the saturated solid solution of four-element compound in phase region Ag_2Te–InTe–$AgIn_2Te_3I$–AgI of the Ag–In–Te–I system. The values of the thermodynamic functions of four-element phase were calculated according in the approximation ΔC_p=const.

4 Conclusions

1. The analytical equation of the temperature dependence of Gibbs energy of the formation from elements of the saturated solid solution of the compound $AgIn_2Te_3I$ in the region $AgIn_2Te_3I$–InTe–Ag_2Te–AgI (I) was obtained.
2. The values of standard integral thermodynamic functions of four-element phase in phase regions (I) were calculated.
3. The symmetry of the lattice of $AgIn_2Te_3I$ compound with the formation of the solid solution is decreased.

References

1. J.W. Perram, *The Physics of superionic conductors and electrode materials* (Springer Science & Business Media, 2013)
2. A.R. West, *Solid state chemistry and its applications* (John Wiley & Sons, 2014)
3. M.V. Moroz, P.Yu. Demchenko, S.V. Prokhorenko, V.M. Moroz, Phys. Solid State. **55**, 1613 (2013).
4. Diffractometer Stoe WinX^{POW}, version 2.21, Stoe & GmbH, Darmstadt (2007)
5. F. Tesfaye, P. Taskinen, M. Aspiala, D. Feng, Intermetallics. **34**, 56 (2013)
6. I. Barin, *Thermochemical data of pure substance* (Weinheim, VCH , 1995)

DOI: 10.1051/epjconf/201713304003

Efficient method for transport simulations in quantum cascade lasers

Mariusz Maczka[1], *Stanislaw* Pawlowski [2]

[1] Department of Electronics Fundamentals, Rzeszow University of Technology, W. Pola 2, 35-959 Rzeszow, Poland
[2] Department of Electrodynamics and Electrical Machine Systems Rzeszow University of Technology, W. Pola 2, 35-959 Rzeszow, Poland

Abstract. An efficient method for simulating quantum transport in quantum cascade lasers is presented. The calculations are performed within a simple approximation inspired by Büttiker probes and based on a finite model for semiconductor superlattices. The formalism of non-equilibrium Green's functions is applied to determine the selected transport parameters in a typical structure of a terahertz laser. Results were compared with those obtained for a infinite model as well as other methods described in literature.

1 Introduction

Quantum cascade lasers (QCLs) are relatively new semiconductor devices [1]-[3]. Nevertheless, they have been successfully used both in scientific research [4-5] and industry applications [6] due to their properties such as high optical power output, wide tuning range of the emitted radiation and room temperature operation. QCLs have proven to be very useful for spectroscopic applications such as remote detection of dangerous gases [7], or pollutants in the atmosphere [8]. QCLs have been also proven effective in medical diagnostics where they are used as breath analyzers [9] or to study plasma chemistry [10]. Scientific studies, both theoretical and experimental, have contributed much to extending the range of their use. Many different approaches can be taken to investigate QCL theoretically. For example, the Monte Carlo method was applied in [11] to predict the current density, population inversion, and optical gain. The method used by other authors [12, 15] to calculate quantitatively QCLs parameters was a non-equilibrium Green's function method (NEGF) operating in a real space. Another known method based on Wannier functions is applied to QCL simulations [16, 17]. Infinite geometrical dimensions of the superlattice structure are assumed and periodic delocalised Bloch functions over the whole structure are considered there. It should be noted, however, that the Bloch theorem was created to describe phenomena occurring in crystal lattices, where the number of cells is incomparable with the number of semiconductor layers in the superlattice heterostructurem which are usually of several dozen, so it is difficult to properly assess the extent to which the assumed infinite dimensions of the model may affect the accuracy of QCL simulations. In [18] the authors compared the results calculated under their model to the results obtained with the finite geometrical dimensions model, and proposed Wannier-function approximation to be used for energies close to the lowest miniband, to demonstrate accuracy of their approach. We showed [19] differences in simulation results obtained with both superlattice models, i.e. with finite and infinite geometrical dimensions, which we called finite (FM) and infinite (IM) model respectively, whereas in [20, 21] we presented a way for constructing the base of quantum states inspired by the Wannier functions base. The base we proposed used the wave functions that were the solution of the Shrödinger equation for the superlattice model of finite geometrical dimensions.

In the present study an efficient method for simulating quantum transport in QCL, based on the finite model is demonstrated. The performance of this method consists of several elements. (i) A quantum states base, inspired by Wannier functions for superlattice properties, is used which allows to limit the Hamiltonian representation solely to the space of energy [22]. For such cases nano-device Hamiltonian matrices are small, so computations run fast. (ii) The scattering process is treated as just a contact described by diagonal matrix with phenomenological parameters η related to the scattering time [23]. (iii) In the calculation scheme where nonequilibrium Green's function equations are solved, a simple approximation inspired by the Büttiker probes common in mesoscopic physics is applied [23-29].

2 Superlattice Model

As mentioned in the previous chapter the finite model for simulating quantum transport in QCL is used. The main elements of this model are shown in Fig.1. It is dedicated to a quantum cascade laser, manufactured on $GaAs/Al_{0.15}Ga_{0.85}As$ superlattice base [30]. The illustrated case represents thermodynamic equilibrium.

* Corresponding author: mmaczka@prz.edu.pl

The model comprises the finite number **n** of **d** length modules. Each module is composed of the layers in the arrangement 7.8 / **2.4** / 6.4 / **3.8** / 14.8 / **2.4** / 9.4 / **5**. Such conducting layers produce a periodic potential, which models the energy scheme of the conduction band - E_C with the difference in values between the quantum well bottom and the energetic barrier height ΔE_C=150 meV

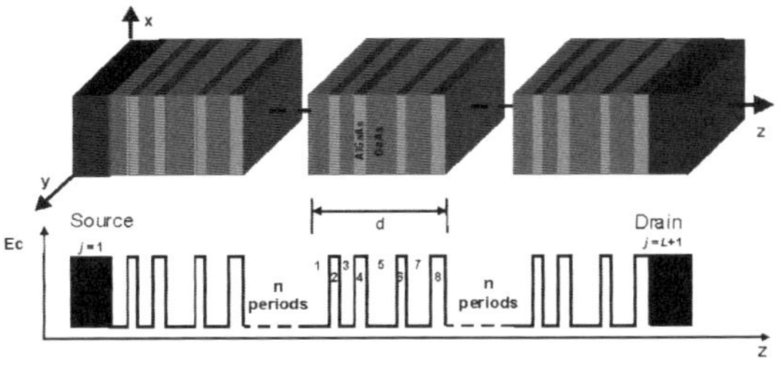

Fig. 1. Schematic model for a cascade laser manufactured on $GaAs/Al_{0.15}Ga_{0.85}As$ superlattice base [30]. The illustrated case represents thermodynamic equilibrium.

Modelling technique for QCL is based on solving the Schrödinger equation for a variable effective mass, namely:

$$\left[-\frac{\hbar^2}{2m_e(z)}\frac{d^2}{dz^2}+V(z)\right]\Psi(z)=E\Psi(z), \quad (1)$$

where $V(z)$ is the potential energy of the electron in the periodic potential resulting from the superlattice semiconductor layers. The solution to equation (1) for the $z_j<z<z_{j+1}$ region of constant potential and constant material composition can be written as wave functions in the form:

$$\psi_j(z)=A_j e^{ik_j(E)(z-z_j)}+B_j e^{-ik_j(E)(z-z_j)}. \quad (2)$$

The functions in the j_{th} layer of the superlattice take on the form of plane counter travelling waves with $k_j=\sqrt{2m_j(E-V_j)/\hbar^2}$, m_j – electron effective mass, E – total energy of an electron, V_j - the potential energy.

The wave functions for the finite superlattice model defined by expression (2) are neither localized, nor periodic [19, 20]. They are grouped within the allowed energy minibands, which are determined by non-zero transmission coefficient value:

$$Tr=\left|\frac{k_{L+1}}{k_1}\left(\frac{A_{L+1}}{A_1}\right)\right|^2, \quad (3)$$

where L stands for the total number of layers in the structure. For expression (3) the assumption that the electron beam concerns only the source ($B_{L+1}=0$) holds and it determines the probability for an electron transition throughout the analyzed structure. The transmission coefficient Tr is equal to 1 for a certain discrete energy value of L-1 number. For the remaining values of energy within the minibands T ranges within $0 < Tr < 1$. Physically it means that the electrons of the same miniband may have different transition probabilities for the structure, and it is believed to have a significant impact on the assessment of its transport properties.

Figure 2.A shows the calculated Tr coefficient values for the wide energy range 0-0.08 eV. Here we see a very narrow energy ranges (minibands) for which this factor has a value close to 1. A more accurate picture of the results we can be seen in Figure 2.B where narrow energy range (0.0620-0.0625 eV) is showed.

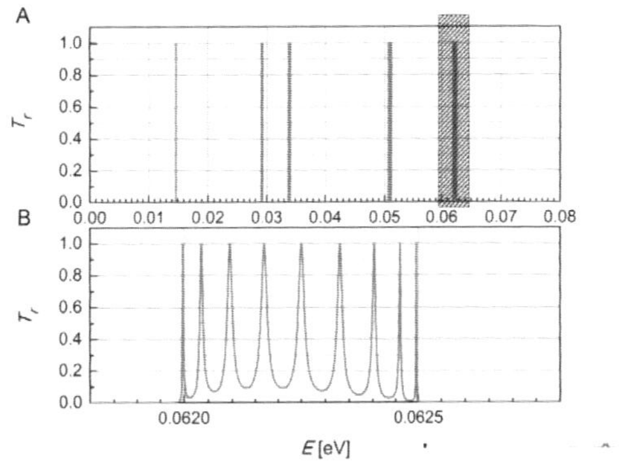

Fig. 2. The calculated values of Tr coefficient for the wide of energy range (0-0.08 eV) A) and the same values for a selected narrow energy range (0.0620-0.0625 eV).

The computation method proposed in this paper uses the base of quantum states inspired by the Wannier functions. In our case the states are maximally localised and defined for a finite model of a nonsymmetrical semiconductor superlattice. These functions were sought according to the formula:

$$V^{\nu}(z)=C\int_{E_{\min}^{\nu}}^{E_{\max}^{\nu}}\psi^{\nu}(E,z)\cdot dE\ , \quad (4)$$

where C is a normalization factor, which shapes the square modulus of the function V^{ν} to the unity of the whole length of the structure, whereas ν indexes the minibands.

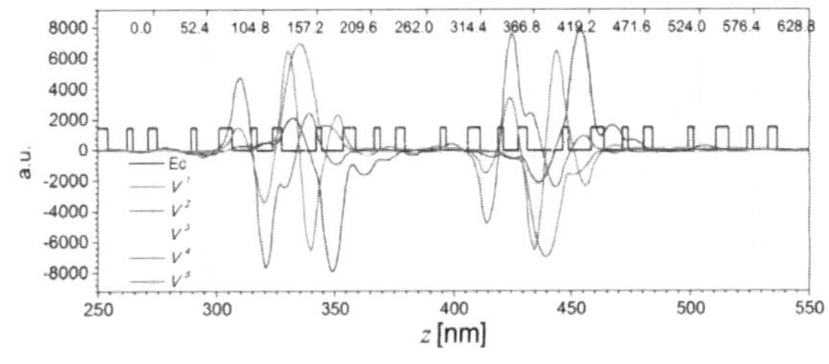

Fig. 3. Maximally localised quantum states representing 5 lowest minibands

It should be noted that $\Psi^{\nu}(E,z)$ functions are not uniquely determined, as their particular embodiment is generally dependent on two complex constants, namely A_1 - the amplitude of the wave travelling from the source and B_{L+1} – amplitude of the wave travelling from the drain. Therefore the search for the maximally localized quantum states in the form of (4) is more difficult than the one where Wannier functions are concerned.

3 The Hamiltonian of the device

Hamiltonian matrix of the device can be written as $\mathbf{H} = \mathbf{H}_0 + \mathbf{H}_R$ where $\mathbf{H}_0$ represents the sum ($\mathbf{H}_{SL} + \mathbf{H}_\xi$) for the cases of unpolarized and polarized superlattices respectively, while the matrix $\mathbf{H}_R$ contains all effects associated with the scattering of electrons. The matrix of $\mathbf{H}_{SL}$ can be written as [22]:

$$\mathbf{H}_{SL} = \sum_{n,\nu}\sum_{\mathbf{k}}\left[E_{\nu\mathbf{k}}\hat{a}^{\nu\dagger}_{\mathrm{n},\mathbf{k}}\hat{a}^{\nu}_{\mathrm{n},\mathbf{k}} + T_1^{\nu}\left(\hat{a}^{\nu\dagger}_{\mathrm{n+1},\mathbf{k}}\hat{a}^{\nu}_{\mathrm{n},\mathbf{k}} + \hat{a}^{\nu\dagger}_{\mathrm{n-1},\mathbf{k}}\hat{a}^{\nu}_{\mathrm{n},\mathbf{k}}\right)\right], \quad (5)$$

where n is the number of the relevant period superlattice of length d, $\mathbf{k}$ is the wave vector, $E_{\nu\mathbf{k}}$ centre of energy for ν miniband, T_1^{ν} represent the off-diagonal couplings between the quantum levels in different periods, whereas parameters $\hat{a}^{\nu\dagger}_{\mathrm{n},\mathbf{k}}$ and $\hat{a}^{\nu}_{\mathrm{n},\mathbf{k}}$ are creation and annihilation states operators, respectively.

The Hamiltonian matrix of the polarized structure $\mathbf{H}_\xi$ takes the form [22]:

$$\mathbf{H}_\xi = \sum_{n,\nu}\sum_{\mathbf{k}}\left\{\begin{array}{l} -e\xi R_0^{\mu\nu}\hat{a}^{\nu\dagger}_{\mathrm{n},\mathbf{k}}\hat{a}^{\nu}_{\mathrm{n},\mathbf{k}} - ne\xi d\delta_{\mu\nu}\hat{a}^{\nu\dagger}_{\mathrm{n},\mathbf{k}}\hat{a}^{\nu}_{\mathrm{n},\mathbf{k}} - \\ +e\xi R_1^{\mu\nu}\left[\hat{a}^{\nu\dagger}_{\mathrm{n+1},\mathbf{k}}\hat{a}^{\nu}_{\mathrm{n},\mathbf{k}} + \hat{a}^{\nu\dagger}_{\mathrm{n-1},\mathbf{k}}\hat{a}^{\nu}_{\mathrm{n},\mathbf{k}}\right]\end{array}\right\}, \quad (6)$$

where e is the electron charge, ξ is the value of the electric field, while variable $R_l^{\mu\nu}$ is calculated according to the formula:

$$R_l^{\mu\nu} = \int dz\, V^{\mu *}(z - ld)zV^{\nu}(z). \quad (7)$$

This relationship takes into account the interaction between both allowed minibands (indices μ, ν) and periods (index l) of the superlattice. The Hamiltonian matrix in our calculations included three periods of the superlattice and five maximally localized quantum states in each of them.

The transformation, described in the previous chapter, from the base of not localized wave functions to the base of maximally localized quantum states allows to reduce the Hamiltonian matrix representation from the energy-position representation to energy representation. This process is also advantageous in reducing the Hamiltonian matrix size (here 15x15), which improves the efficiency of the presented method.

In NEGF theory scattering processes expressed in $\mathbf{H}_R$ are applied in the form of self-energies as described in the next section.

4 Green's functions and self energy approach

For calculating transport characteristics in QCL it is necessary to use nonequilibrium, i.e. retarded G^R and correlation $G^<$, Green's functions to define the nonequilibrium stationary state of the system. NEGF theory contains a description of their use for that purpose, and a self-consistent Born approximation [16] remains to be a quite popular calculation method. In this paper we propose an approach based on other methods to determine Green's functions. We were inspired by Büttiker probes (see Fig. 4), widely used in mesoscopic physics, where the parameter $\Gamma_b = i\left(\Sigma_b - \Sigma_b^+\right)$ [27] describing the exchange of electrons between the quantum state of each module b and every reservoir of electrons in active device, taking into account the considered scattering effects, are introduced. These effects are included in the self-energy matrix Σ.

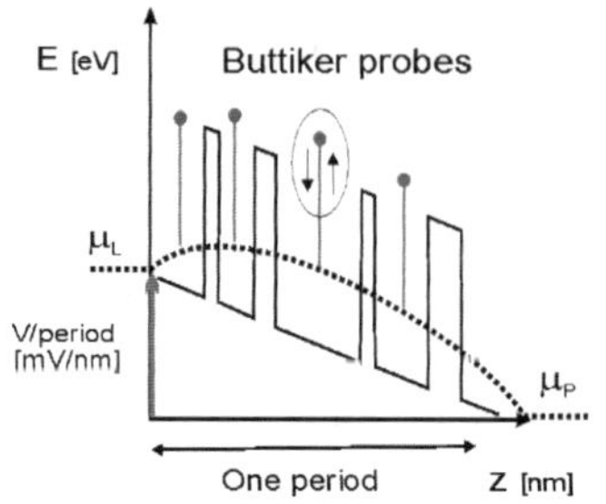

Fig. 4. The profile of a generic mode is illustrated along with the placement of Büttiker probes.

For such an approach the matrix of the spectral function can be written in the form [27]:

$$\mathbf{A}_b^N = \mathbf{G}_b^R \mathbf{\Gamma}_b^N \mathbf{G}_b^{R+}, \quad (8)$$

where N indexes all electron reservoirs. It should be mentioned here that diagonal elements of the matrix $\mathbf{A}$ give information about the local density of states defined as [27]:

$$\mathbf{D}_b = \frac{1}{2\pi}\left(\mathbf{A}_b^N\right) = \frac{1}{2\pi}\left|\mathbf{G}_b^R\right|^2 \mathbf{\Gamma}_b^N. \quad (9)$$

The transmission coefficient of electrons between the two reservoirs of m and n is determined with the following formula [27]:

$$\mathbf{T}_b^{mn} = \mathrm{Trace}\left[\mathbf{\Gamma}_b^m \mathbf{G}_b^R \mathbf{\Gamma}_b^n \mathbf{G}_b^{R+}\right] = \mathbf{\Gamma}_b^m \left|\mathbf{G}_b^R\right|^2 \mathbf{\Gamma}_b^n. \quad (10)$$

With the known local density of states, and scattering effects taken into account by the factor $\mathbf{\Gamma}$, we can define the local concentration of electrons as [27]:

$$n_b = \frac{1}{\hbar}\sqrt{\frac{m^*_{e,b}k_BT}{2\pi^3}}\sum_N\left(\mathbf{A}_b^N\right)\times F_{-1/2}\left(\mu_N - E_z\right) \quad (11)$$

where F is the function of distribution of electrons, determined by the Fermi levels μ_N calculated for Büttiker probes and normalized by a factor k_BT. Similarly, the current flowing through the probe to the quantum state b may be expressed as [27]:

$$I_{sB} = \frac{e}{\hbar^2}\sum_b\sqrt{\frac{m^*_{e,b}k_BT}{2\pi^3}}\sum_N \mathbf{T}_b^{mn}\left[F_{-1/2}\left(\mu_m - E_z\right) - F_{-1/2}\left(\mu_n - E_z\right)\right], \quad (12)$$

where the Fermi levels μ_n and μ_m of the Büttiker probes are determined from the condition of current continuity:

$$I_{sB} = \int_{-\infty}^{\infty} I_{sB}(E_z)dE_z = 0 \quad (13)$$

DOI: 10.1051/epjconf/201713304003

This leads to an iterative equations [27]:

$$\mathbf{J}^{mn} = \frac{\partial I_{sB}}{\partial \mu_m} = \frac{e}{\hbar^2}\int_{-\infty}^{\infty}\sum_b \sqrt{\frac{m_{e,b}^{*} k_B T}{\pi^3}} \times \left\{ \frac{\partial F_{-1/2}(\mu_m - E_z)}{\partial \mu_m} \sum_n \mathbf{T}_b^{mn} \right\} dE_z \, , \tag{14}$$

$$\mathbf{J}^{mn} = \frac{\partial I_{sB}}{\partial \mu_n} = \frac{e}{\hbar^2}\int_{-\infty}^{\infty}\sum_b \sqrt{\frac{m_{e,b}^{*} k_B T}{\pi^3}} \times \left\{ \mathbf{T}_b^{mn} \frac{\partial F_{-1/2}(\mu_n - E_z)}{\partial \mu_n} \right\} dE_z \, , \tag{15}$$

and correction of the chemical potential at each iteration of each of the Büttiker probes in accordance with the relationship:

$$\Delta\mu^{sB} = -\mathbf{J}^{-1} I_{sB} \tag{16}$$

until no change of μ_{sB} occurs. Having the chemical potential on the Büttiker probes, we can specify self energies $\mathbf{\Sigma}^{\mathbf{R}}$ and $\mathbf{\Sigma}^{<}$ as diagonal matrices with elements in the form $i\eta$ and $i\eta f_{Bt}(E)$, respectively, where $f_{Bt}(E) = 1/[\exp(E - \mu_{zB}/k_B T) - 1]$, while η is a phenomenological parameter related to the scattering time ($\eta = \hbar/\tau$) [22]. The next step is to solve two kinetic equations. The first of them is the Dyson equation [23]:

$$(E\,\mathbf{I} - \mathbf{H} - \mathbf{\Sigma}^{R})\mathbf{G}^{R} = \mathbf{I} \, , \tag{17}$$

where $\mathbf{G}^{R}$ is the matrix of the retarded Green's functions, which can be define as [24]:

$$G^{R}(E) = \lim_{\eta \to 0^{+}} G(Z = E + i\eta) \tag{18}$$

The second is the Keldysh equation [31]:

$$[\mathrm{G}^{<}] = [\mathbf{G}^{R}][\Sigma^{<}][\mathbf{G}^{R}]^{\dagger} \, , \tag{19}$$

which allows to determine the correlation function $G^{<}$.

It is worth emphasizing that in the method described here, the kinetic equations (17) and (19) need to be solved only once, unlike in the Born approximation method, where they are solved repeatedly in the self-consistent process together with the equations for the self energies.

According to the assumptions made here, the density of current flowing through the laser structure, can be written just as it was for the description of the Hamiltonian as $J = J_0 + J_R$ [16], where J_0 is the current with the effects of electron scattering excluded, while J_R is a component taking then into account. As shown by the authors in [32] for the structure under consideration the J_R component disappears, and thus the total current density can be approximated by the J_0 defined as [32]:

$$J_o = \frac{2e}{\hbar\vartheta}\sum_{\alpha,\beta,\mathbf{k}}\int \frac{dE}{2\pi}\left[\hat{H}_0, \hat{z}\right]_{\alpha,\beta} G^{<}_{\alpha,\beta,\mathbf{k}}(E) \tag{20}$$

where the matrix $\left[\hat{H}_0, \hat{z}\right]$ is formed in the base of maximally localized quantum states variable ϑ is the volume of polarized structure. By this relationship the density of current were calculated and the obtained results are described in the next section.

5 Calculations of Green's functions

The model described in the previous chapters produced the simulations results illustrated below. In Figure 5 (left side) the imaginary diagonal elements of the matrixes $\mathbf{G}^{\mathbf{R}}$ and $\mathbf{G}^{<}$ are plotted for wave vector $\mathbf{k}$=0. There, the respective density of states ($N_{1D}(E)$) and the occupation of these states by electrons ($n_{1D}(E)$) are illustrated. The simulations were carried out with the eclectic field of 50 mV/period applied. The Fig. 5 (right side) contains also, the results of the Green's functions calculations, using the infinite model of superlattice [33]. This juxtaposition shows differences in distribution of peak amplitudes of Green's functions. This is a consequence of different boundary conditions used in both models.

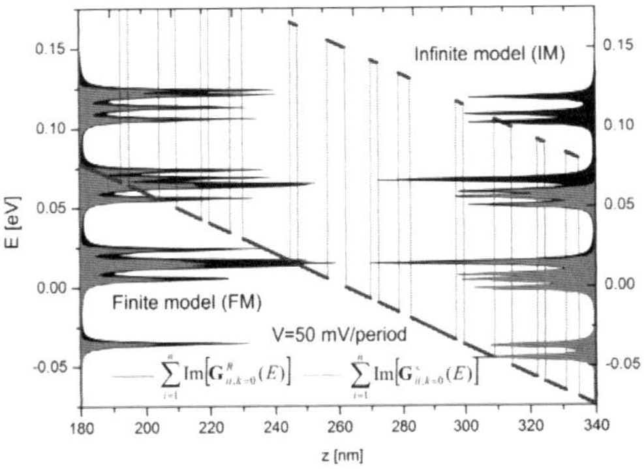

Fig. 5. Examples of the density of states functions $N_{1D}(E)$ and occupation functions $n_{1D}(E)$, for $\mathbf{k}$=0 calculated by use Büttiker probes, and the eclectic field of 50 mV/period applied. The simulations were carried out for the finite and the infinite models of superlattice.

By summing up one dimensional density of states functions for all states over the selected range for the wave vector $\mathbf{k}$, the function of the density of states for the three geometric dimensions (3DOS) is obtained. This function can be written as $N_{3DOS}(E) = \sum_{\nu,\mathbf{k}} \mathrm{Im}\left[\mathbf{G}^{R}_{ii}(E)\right]$.

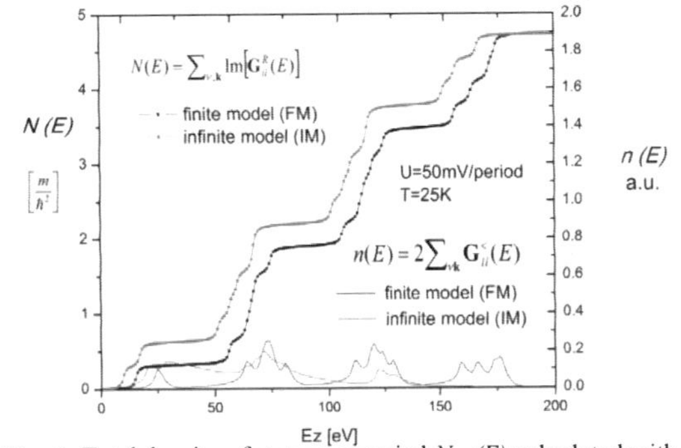

Fig. 6. Total density of states per period $N_{3D}(E)$ calculated with FM, and occupation functions per period $n_{3D}(E)$ calculated with Büttiker probes, and the eclectic field of 50 mV/period applied. The simulations were carried out for the finite and the infinite models of superlattice.simulations were carried out for the finite and the infinite models of superlattice.

 DOI: 10.1051/epjconf/201713304003

In a similar way we obtained the occupation function ($n_{3DOS}(E)$) for the three geometric dimensions $n_{3DOS}(E) = 2\sum_{\mathbf{k}} \mathbf{G}^{<}_{ii}(E)$. Figure 6 shows graphs of $N_{3DOS}(E)$ and $n_{3DOS}(E)$ functions with the electric field of 50 mV/period applied. There are the results for the finite and the infinite models of superlattice.

The maps shown in Figure 7 are an extension of the results p resented in Figures 5 and 6. The density of states (Fig. 7A) and the occupation functions (Fig. 7B), defined as $N(E,k_{\parallel}) = \frac{1}{\pi}\sum_{v} \mathrm{Im}(\mathrm{Tr}([G^{R}]))$ and $n(E,k_{\parallel}) = \frac{2}{\pi}\sum_{v} \mathrm{Im}(\mathrm{Tr}([G^{<}]))$ respectively, are illustrated. All these functions are shown for Büttiker probes approach, with the bias voltage of 50 mV/period applied. The results are plotted for the finite and the infinite models of superlattice, in a way that allows them to compare.

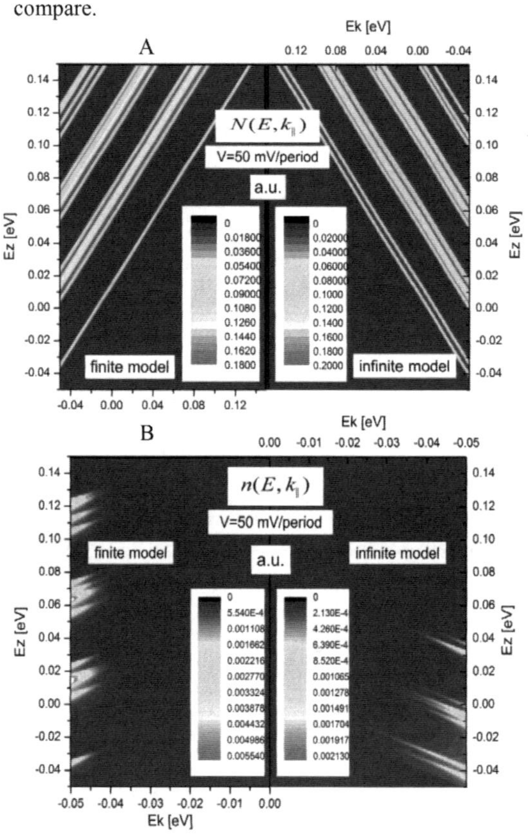

Fig.7. Density of states *N(E,k||)* (part A) and occupation functions *n(E,k||)* (part B), calculated for the bias voltage of 50 mV/period with Büttiker probes by use finite and infinite models of supperlattice.

The method presented is advantageous in terms of the effectiveness of simulating U-I characteristic for QCLs. An example illustrated in Fig. 8, where the current density Jo for FM and IM models was calculated according to (20) and versus voltage polarity per period is plotted. The calculations were conducted for a superlattice temperature of 25 K [30].

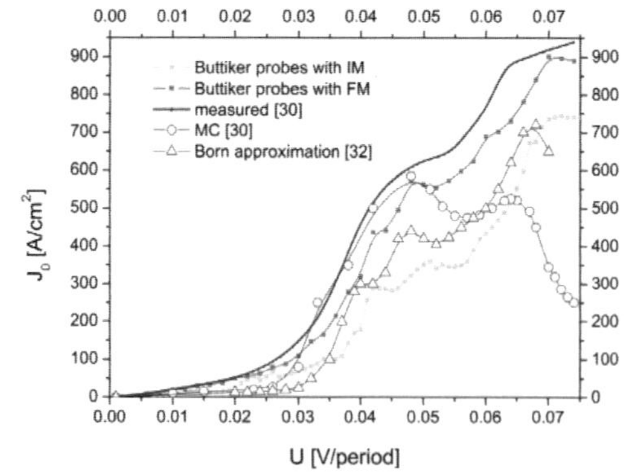

Fig. 4. Current-voltage characteristics of the QCL structure [30] obtained for different approaches. The blue solid line plotted the measured current-voltage characteristics of the studied structure..

The obtained characteristics show similarity to the results presented in [32], which were calculated using the Born approximation and in [30] where we can find the measured and simulated (with Monte-Carlo (MC) method) current-voltage characteristics for the structure considered here. It seems to confirm applicability of Büttiker probes for simulating terahertz lasers with the superlattice finite model.

Summary

The theory of Büttiker probes applied to efficient calculations of the transport parameters for quantum cascade lasers, based on the finite model of superlattice, was described.

The presented simulation proved the proposed approach to have given results in line with other applicable methods described in the literature. Superiority of the developed procedure for finding laser parameters as being much faster and easier to operate in comparison to the other methods reported elsewhere was demonstrated. It gives ground for further research into the proposed approach.

References

1. J. Faist, F. Capasso, D. L. Sivco, A. L. Hutchinson, S-N. G. Chu, A. Y. Cho, Appl. Phys. Lett. **72**, 680-684 (1998)
2. R. Colombelli, F. Capasso, C. Gmachl, A. L. Hutchinson, D. L. Sivco, A. Tredicucci, M. C. Wanke, A. M. Sergent, A. Y. Cho, Appl. Phys. Lett. **78,** 2620 – 2622 (2001)
3. K. Kosiel, M. Bugajski, A. Szerling, J. Kubacka-Traczyk, P. Karbownik, E. Pruszyńska-Karbownik, J. Muszalski, A. Łaszcz, P. Romanowski, M. Wasiak, W. Nakwaski, I. Makarowa, P. Perlin, Photonics Letters of Poland **1**, 1,16-18, (2009)
4. M. Bugajski, K. Pierściński, D. Pierścińska, A. Szerling, K. Kosiel, Photonics Letters of Poland **5**, 3, 85-87 (2013)

DOI: 10.1051/epjconf/201713304003

5. P. Karbownik, A. Trajnerowicz, A. Szerling, A. Wójcik-Jedlińska, M. Wasiak, E. Pruszyńska-Karbownik, K. Kosiel, I. Gronowska, R.P. Sarzała, M. Bugajski, Optical and Quantum Electronics **47**, 4, 893-899 (2015)
6. E. Normand; I. Howieson; M. T. McCulloch, Laser Focus World **43** 4, 90–92 (April 2007)
7. Z. Bielecki, J. Wojtas, T. Stacewicz, Elektronika-Konstrukcje-Technologie-Zastosowania, 11, 44 (2014)
8. M. Miczuga, K. Kopczyński, J. Pietrzak, Elektronika-Konstrukcje-Technologie-Zastosowania, 11, 47 (2014)
9. M. Hannemann, A. Antufjew, K. Borgmann, F. Hempel, T. Ittermann, S Welzel., K.D. Weltmann, H. Völzke, J. Röpcke , Journal of Breath Research, **5** ,027101 (2011)
10. N. Lang; J. Röpcke; S. Wege; A. Steinach, Eur. Phys. J. Appl. Phys. **49**, 13110:3 (2009)
11. E. Bellotti, K. Driscoll, T. D. Moustakas and R. Paiella, Journal of Applied Physics **105**, 11 (2009)
12. S. Birner, T. Zibold, T. Andlauer, T. Kubis, M. Sabathil, A. Trellakis, P. Vogl, IEEE Transactions on Electron Devices **54**, 9, 2137 (2007)
13. T. Kubis, C. Yeh, and P. Vogl, Phys. Rev. B **79**, 19, 195323-1–195323-10 (2009)
14. G. Hałdaś, A. Kolek, and I. Tralle, Journal of Quantum Electronics **47**, 6 (2011)
15. A. Kolek, G. Hałdaś, M. Bugajski, Appl. Phys. Lett. **101,** 061110 (2012)
16. S.-C. Lee and A. Wacker, Phys. Rev. B **66**, 245314 (2002)
17. A. Wacker, S.-C. Lee, M.F. Pereira Jr, Advances in Solid State Physics ed. by B. Kramer **43**, Springer, Berlin, 369–380 (2003)
18. A. Wacker and B. Y.-K. Hu, Phys. Rev. B **60**, 23, 16039-16049 (1999)
19. M. Mączka, S. Pawłowski, J. Plewako, Electrical Review 8, 93 (2011)
20. S. Pawłowski, M. Mączka, Przegląd Elektrotechniczny 11, 322-327 (2013)
21. M. Mączka, S. Pawłowski, Przegląd Elektrotechniczny 12, 245 (2013)
22. A. Wacker, Physics Reports **357**, 1-111 (2002)
23. S. Datta, Phys. Rev. B **40**, 5830 (1989)
24. S. Datta, Electronic transport in mesoscopic systems (*Cambridge University Press*, 1995)
25. M Büttiker, Phys. Rev. Lett. **57**, 1761 (1986)
26. Texier and M. Büttiker, Phys. Rev. B **62**, 7454 (2000)
27. R.Venugopal, M. Paulsson, S. Goasguen, S. Datta, and M. Lundstrom, J. Appl. Phys. **93**, 5613 (2003)
28. P. Greck, S. Birner, B. Huber, P. Vog, Optics Express **23**, 6587-6600 (2015)
29. M. Mączka, Przegląd Elektrotechniczny, 7, 190-196 (2016)
30. H. Callebaut and Hu Q., J. Appl. Phys. **98**, 104505 (2005)
31. L. V. Keldysh, Sov. Phys. JETP **20**, 4, 1018–1026 (1965)
32. S.-C. Lee, F. Banit, M. Woerner, and A. Wacker, Phys. Rev. B **73**, 245320 (2006)
33. M. Mączka, S. Pawłowski, Elektronika-Konstrukcje-Technologie-Zastosowania, 11, 187 (2014)

Surface-enhanced Raman scattering and Plasmon effect for enzymatic bionanocomplexes characterization

Renata Wojnarowska-Nowak[1], *Jacek* Polit[1], *Daniel* Broda[2], *Yaroslaw* Bobitski[3], *Zbigniew* Starowicz[4], *Mykhailo* Gonchar[2,5], and *E. M.* Sheregii[1]

[1]Centre for Microelectronics and Nanotechnology, University of Rzeszow, S. Pigonia 1, 35-959 Rzeszow, Poland
[2] Department of Biotechnology, Institute of Applied Biotechnology and Basic Sciences, University of Rzeszow, Werynia 502, 36-100 Kolbuszowa, Poland
[3] Faculty of Mathematics and Natural Sciences, University of Rzeszow, S. Pigonia 1, 35-959 Rzeszow, Poland
[4] Institute of Metallurgy and Materials Science Polish Academy of Sciences, Reymonta 25, 30-059 Krakow, Poland
[5] Institute of Cell Biology, National Academy of Sciences of Ukraine, Drahomanov 14/16, Lviv 79005, Ukraine

Abstract. Cholesterol oxidase (ChOX) enzyme is one of the most important analytical enzyme, used for cholesterol assay in clinical diagnostics as well as food production, and the developing of innovative solutions for improving the selectivity and accuracy of the analysis including bio-nanotechnological approaches is still ongoing. The Surface Plazmon Resonance (SPR) and the surface enhanced Raman scattering (SERS) as specific for nanocurriers effects were observed what enable us to research the oscillation spectra of the ChOX enzyme. The vibrational lines are attributed to chemical functional groups existing in enzyme, for example, amino acids, amide groups as well as for cofactor. For the improving the SERS effect the gold nanoparticles – ChOX bionanocomplexes were analyzed in combination with gold-coating gratings as a promising plazmonic material.

1 Introduction

In the past few years surface enhanced Raman scattering (SERS) spectroscopy has been widely employed as a powerful and extremely sensitive technique [1]. Due to high sensitivity, selectivity and non-destructive data acquisition the method has been applied in chemistry and biochemistry [2]. Because of possibility to detect small amount of substances, or even single molecules, it can be employed successfully for analitical purposes [3]. SERS effect have been used in biological and medical applications, as well as environmental monitoring, through the development of optical biosensors [4,5].

SERS signal can be promoted by means of surface plasmon resonance (SPR) and localized surface Plasmon resonances (LSPRs) [6,7]. Surface Plasmon's (SPs) are optical waves that originate from coupled collective oscillations of the electron plasma and associated electromagnetic field on metallic surface [1]. The SPR and LSPR phenomena can occur when chemical or biological elements interact with noble metal nanoparticles or nanostructures. The enhancing level is dependent on the composition, shape and size [8] of nanoparticles or nanostructures, as well as their geometry and roughens [9]. Occurrence of a strong SERS effect is also highly dependent on the interaction between investigated molecules and the surface of plasmonic nanostructures, which must be in close distance [7]. Therefore, it is important to develop the substrates which can get the maximum enchancing of the registered signal, while their production and using does not carry any major technical problems.

In presented work two kind of substrates were emplyed to obtained the plasmon effect: the gold nanoparticles (AuNPs) and polycarbonate gratings coated with gold, as a promising plasmonic materials [5,10]. The ChOX-AuNP bionanocomplex was situated on the grating which have improved optical properties and enhance the SERS effect additionally. The phenomenon of surface enhanced Raman scattering has been observed and has been used for the analysis of ChOX enzyme.

2 Experiment

2.1 Sample preparation

For the SERS study a bionanocomplex of cholesterol oxidase enzyme (ChOX, EC 1.1.3.6) and AuNP was performed [5]. The ChOX-AuNP bionanocomplexes were prepared by covalently binding of the gold nanoparticles and enzym via organic linker, as described in Ref. 5. AuNPs were obtained by modyfied Turkevich method with using chloroauric acid ($HAuCl_4$) and sodium citrate ($Na_3C_6H_5O_7$) as a substrates of the reaction. On the surface of AuNP a mercaptohexa-

* Corresponding author: rwojnar@ur.edu.pl

 DOI: 10.1051/epjconf/201713305001

decanoic acid (MHDA) was immobilized as a linker for the ChOX enzyme binding.

In second step of presented study the prepared bionanocomplex was placed on the metal diffraction gratings as a plasmonic surfaces. They were made with using hologram replication technology [11] on the thin polycarbonate film. Grating depth is 300 nm, with various grating period from 400 nm to 10000 nm. The surface was covered by gold layer with 120 nm of layer thickness, by using vacuum deposition method.

2.2 Methods of characterization

The main methods for the ANPs and diffraction gratings characterization were two microscopic techniques: Atomic Force Microscopy (AFM, with using Innova Bruker microscope, in Tapping Mode), and Scanning Electron Microscopy (PhenomPro microscope).

The SERS effect was studied with using two spectrometers with different excitation sources: Smart Raman DXR Thermo Scientific spectrometer, with a 780 nm excitation wavelength (14 mW laser power), and InVia Micro Raman Renishaw spectrometer, with a 633 nm excitation (17 mW laser power).

3 Results and discussion

The AFM image of the AuNPs is shown in Fig. 1. The nanoparticles have spherical shape and size at about 45 nm, however they are not completely homogenous. It is very difficult to obtain only one size of nanoparticles by the colloidal synthesis method. The size of AuNPs is dependent on the concentration of substrate and reducing agent in the synthesis mixture [5] as well as reaction conditions. After conjugation with enzyme the AuNPs are more aggregate and in various shapes but mainly spherical. It is seen in insert the assuming distribution of a near field for single spherical nanoparticle.

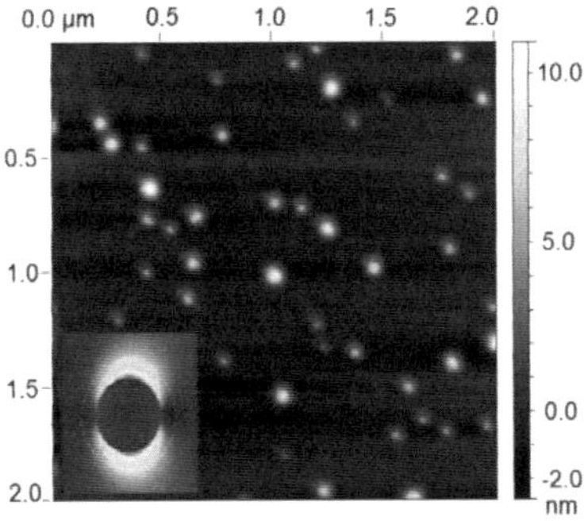

Fig 1. AFM imageing of AuNPs used for the synthesis of the bionanocomplex and distribution of a near field for single spherical nanoparticle (insert).

Topography of the prepared gold-coated polycarbonate grating is presented in Fig. 2. The grating period is determined to be 400 nm. The groove depth after covering by gold layer is estimated to be about 120 nm.

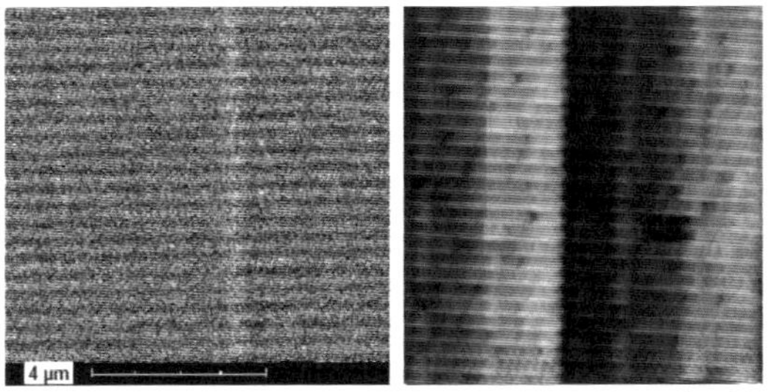

Fig. 2. SEM image (left) and AFM image (right) of gold-coated polycarbonate grating.

As was mentioned above ChOX enzyme is covalently immobilized with the AuNPs by MHDA linker and this connection is schematically shown in Fig. 3. The connection between the AuNPs and the MHDA linker is possible by presence of S-H thiol group and Ag-S bond formation [8]. Carboxyl group of linker enables the protein binding by forming a peptide bond with the specified amino acids.

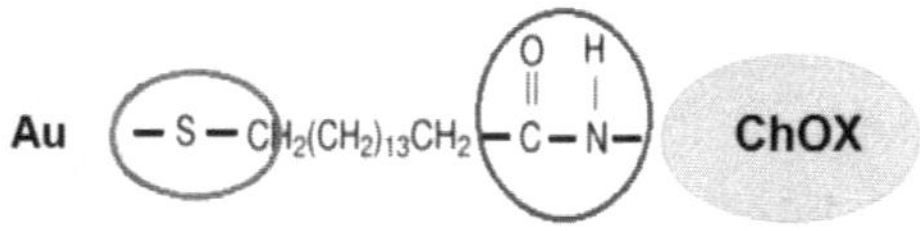

Fig. 3. Scheme of bionanocomplex consisting with AuNP, MHDA linker and cholesterol oxidase enzyme, with marked chemical bonds generated between them.

By this manner, the ChOX as an important analytical enzymes was investigated in conducted study. ChOX is used for the determination of the serum cholesterol level in clinical diagnostics [12,13]. It is an important biomarker of coronary heart disease, arteriosclerosis, and lipid metabolism dysfunction [14]. Furthermore, high concentration of cholesterol in food has a negative impact on our health, and because of this cholesterol level should be monitored in a food production [15]. Besides that ChOX is a promissing novel antibacterial therapeutic agent and herbicide [15]. Therefore, it is important also to develop convenient methods of controlling the presence of the enzyme.

The SERS effect has been observed for both of the prepared two bionanosystems. In Fig. 3 the Raman spectrum of the AuNP-ChOX bionanocomplex is shown (780 nm laser source). In Fig. 4 the Raman spectrum of bionanocomplex placed on the surface of gold-coated grating is presented (633 nm laser source).

Different degrees of the signal amplification (SERS) for prepared systems were achieved, which allows an optimization the architecture of prepared substrates. With using of gold-coating gratings with presented above parameters registered Raman signal was maximal.

Obtained Raman spectra can be used for the identification of functional groups existing in the ChOX enzyme [5]. Analogically to spectra obtained in Ref. 5,

DOI: 10.1051/epjconf/201713305001

the most intense line observed at 1530 cm^{-1} in the Fig. 4. is assigned with the chemical bonds characteristic for proteins: N-H, C-N, and N-C-O vibrations, named as amide II [16,17].

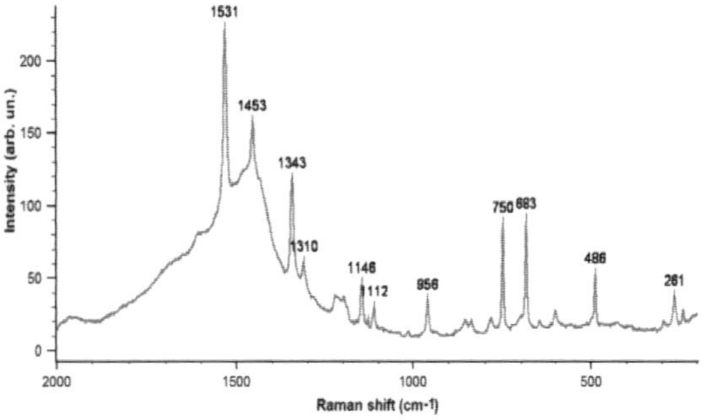

Fig. 4. SERS spectrum of AuNP-ChOX bionanocomplex, under 780 nm excitation wavelength.

Other bands characteristic for proteins are also observed. Amide I and III lines are recorded at 1610 cm^{-1} [16] and at around 1220 cm^{-1} [18], however they are significantly weaker. Other lines at 1146 cm^{-1}, 800-900 cm^{-1} and 750 cm^{-1} can be attributed to the specific amino acids [19,20,21]. Some of the identified lines are not specific as a line at 1453 cm^{-1} associated witch CH vibration existing in linker and in protein structure as well [20]. Besides typical functional groups of proteins the vibration lines attributed to the Flavin adenine dinucleotide (FAD) cofactor have been observed (such as the adenine ring at 1343 and 1310 cm^{-1} [20], P-O around 1200 cm^{-1} and C=C binding at 1112 cm^{-1}) [18].

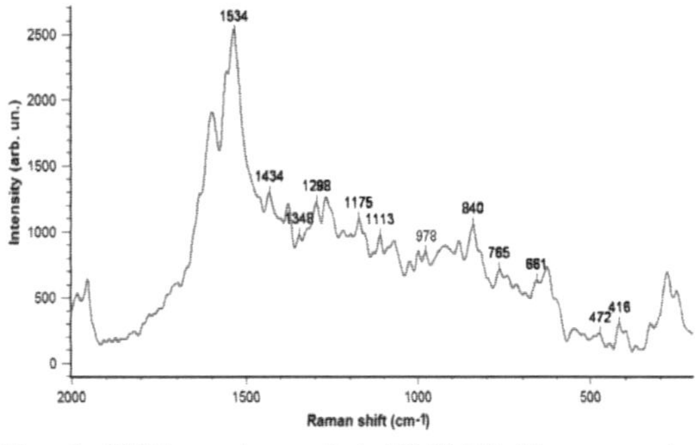

Fig. 5. SERS spectrum of AuNP-ChOX bionanocomplex placed on the gold grating, under 633 nm excitation wavelength.

The same characteristic lines can be observed in Fig. 5. However, the spectra obtained with investigated substrates are slightly modified. Some shifts of the bands, weakening of lines and strengthen others are observed. It can be cased by differen hotspot places in modyfied plasmonic system in combination with a high complication of protein molecule. The most intense line at 1531 cm^{-1}, is slightly shifted to 1534 cm^{-1}. Other lines attributed to amide I, amide II, some amino acids and FAD cofactor are recorded also, and presented in Table 1.

Table 2. Identification of the lines observed in the Raman spectrum presented in Fig. 5.

	Position of line (cm^{-1})	Identification	Reference
1	1602	C=O band	16,17
2	1534	N-H, C-N, N-C-O bands	16,17
3	1434	C-H str. vibration	17,20
4	1349	Adenine, ring I of FAD	20
5	1298	C-N and N-H band	16,17
6	1264	ring II of FAD molecules	22
7	1220	C-N str., N-H	18
8	1113	C=C of FAD	18
9	926	C-C band	17
10	800-900	arginine amino acid, C-H out of plane	19,20,21
11	765	tryptophan	21

The resulting SERS effect suggests that bionanocomplexes AuNP-enzyme can be used for analytical purposes, and an optimisation of additional substrates to enchance the signal should be conducted.

Conclusion

The phenomenon of SERS effect was obtained using additional substrate of the gold covered diffraction grating. Prepared plasmonic surfaces could be usable SERS-platform for study of enzymes. Obtained SERS spectra can be used for the identification of functional groups existing in the examined enzyme or other biological molecules. Different degrees of the signal amplification (SERS) for prepared systems were achieved, which allows an optimization the architecture of prepared substrates. The gold-coated diffraction gratings in combination with a gold nanoparticles-enzyme bionanocomplexes can be perspective analytical instrument.

References

1. K.A. Willets, R.P. Van Duyne, Annu Rev Phys Chem. **58**, 267 (2007)
2. S.Nie, S.R. Emory, Science **21**, 275 (5303), 1102 (1997)
3. D.A. Stuart, J.M. Yuen, N. Shah, O. Lyandres, C.R. Yonzon, M.R. Glucksberg, J.T. Walsh, R.P. Van Duyne, Anal. Chem. **78** (20), 7211 (2006)
4. N.A. Mungroo, S. Neethirajan, Biosensors **21**, 4, 472 (2014)
5. R. Wojnarowska, J. Polit, D. Broda, M. Gonchar, E. M. Sheregii, Appl. Phys. Let. **106**, 103701 (2015)
6. H. Chen, X. Kou, Z. Yang, W. Ni, J. Wang, Langmuir **20**, 24(10), 5233 (2008)
7. J. Ye, F. Wen, H. Sobhani, J.B. Lassiter, P. Van Dorpe, P. Nordlander, N.J. Halas, Nano Lett. **12**, 3, 1660 (2012)

8. R. Wojnarowska, J. Polit, D. Broda, M. Gonchar, E. M. Sheregii, AMM **60,** 3, 2289 (2015)
9. M.C. Estevez, M.A. Otte, B. Sepulveda, L.M. Lechuga, Anal. Chim. Acta. **806,** 55 (2014)
10. B.K. Singh, A.C. Hillier, Anal. Chem. **80** (10), 3803 (2008)
11. M.T. Gale, Journal of imaging science and technology **41**, 3, 211 (1997)
12. U. Saxena, M. Chakraborty, P. Goswami, Biosens. Bioelectron. **26**, 3037 (2011)
13. L. Pollegioni, L. Piubelli, G. Molla, FEBS Journal, **23**, 1742 (2009)
14. P. Srisawasdi, P. Jearanaikoon, N. Wetprasit, B. Sriwanthana, M.H. Kroll, P.H. Lolekha, Clin Chim Acta. **372**, 103 (2006)
15. A. Vrielink, Subcell. Biochem. **51**, 137 (2010)
16. J. Gelder, K. De Gussem, P. Vandenabeele, L. Moens, J. Raman Spectrosc. **38**, 1133 (2007)
17. Rygula, A., Majzner, K., Marzec, K. M., Kaczor, A., Pilarczyk, M., Baranska, M., J. Raman Spectrosc. **44,** 1061 (2013)
18. R. F. Brandao, R. L. Quirino, V. M. Mello, A. P. Tavares, A. C. Peres, F. Guinhos, J. C. Rubim, P. A. Z. Suarez, J. Braz. Chem. Soc. **20,** 5, 954 (2009)
19. N. A. Atamas, A. M. Yaremko, T. Seeger, A. Leipertz, A. Bienko, Z. Latajka, H. Ratajczak, A. J. Barnes, J. Mol. Struct. **708,** 189 (2004)
20. R. Tuma, J. Raman Spectrosc. **36**, 307 (2005)
21. A. Brambilla, A. Philippidis, A. Nevin, D. Comelli, G. Valentini,D.Anglos, J. Mol. Struct. **1044**, 121 (2013)
22. Zheng, Y., Care, P. R., Palfey, B. A., J. Raman Spectrosc. **35**, 521 (2004)

Influence of calcium blockers on the SPR of erythrocytes

Olena V. Shynkarenko[1], *Orest* Tril[2], *Renata* Wojnarowska[3], *Sergiy* Prohorenko[3,4] and *E. M.* Shergii[3]

[1] Lashkaryov Institute of Semiconductor Physics, NAS of Ukraine. Kyiv, Ukraine
[1] Lviv State Oncology Regional Treatment and Diagnostic Centre. Lviv, Ukraine
[3] Center for Microelectronics and Nanotechnology, University of Rzeszów. Rzeszów, Poland
[4] Lviv Polytechnic National University. Lviv, Ukraine

Abstract. One of the promising areas of research is the impact of calcium channel blockers (CB) of biological fluids. This paper shows that the CB impact on a biological fluid can be efficiently combine with the surface plasmon resonance (SPR). It is shown that the addition of CB at the SPR measurements affect the stability of membranes and acts differently on the kinetics of erythrocytes ligament in the different groups of people.

1 Introduction

The structural and functional properties of red blood cells are mostly determined by rates of rigidity and elasticity of the membrane and by adhesive proteins on its surface [1]. High sensitivity of erythrocytes to the most of pathological changes in many diseases causes great diagnostic significance using a monitoring of the degree of elasticity of erythrocytes [2-4].

The main methods that are widely used in clinical practice to determine erythrocytes elasticity are method of filtering, micro-pipette method, method of centrifugation and others [4]. Nevertheless, they are quite labour-intensive and time-consuming measurements. For practical purposes it is important to have rapid research methods of red blood cell mass that use minimum quantities of biofluid samples. The ability to test the response of the red blood cell to the stress conditions (hypotonic or hypertonic solution, medicine etc.) is relevant. To solve these tasks the methods of diffractometric and the use of laser-traps [5,6] are being developed. Unfortunately, they have important shortcomings: the method of laser-traps has a high error level of measurement – 19-41% [5,7]. The present work proposes an effective method of the properties characterization of the red blood cells (RBC) using the surface plasmon resonance (SPR) measurements. We use calcium blockers [3] influence on the erythrocyte mass to demonstrate that SPR method can be effectively used to study effects of medicines on properties of the RBC.

2 Experiment

2.1 Role of calcium blockers

The widespread implementation of calcium blockers (CB) into clinical practice has led to a detailed study of calcium homeostasis [3].

It was found that ionized Ca2 + is involved in the regulation of the most of intracellular processes. Investigation of the CB influence on human erythrocytes makes it possible to obtain new fundamental knowledge concerning the organism as a whole and can be used in the future for diagnosing diseases. Properties of human erythrocytes are changed over time and are influenced by various factors.

To assess the stability of erythrocyte membranes, showing the condition of the organism as a whole, such methods as Definition of osmotic resistance of erythrocytes, Evaluation of peroxide resistance of erythrocytes, Determination of permeability parameters of the erythrocyte membranes using the method of urea hemolysis have been used [8-11]. These methods require considerable labor inputs and a long time to complete investigations. The SPR based sensors was used to perform the measurements in the current work. For above purpose highly specific proteins that bind complementary proteins from a liquid sample are immobilized on the working surface. On the basis of known investigated antigen-antibody pairs the vast majority of human blood tests for presence of antibodies to certain diseases are carried out by ELISA (Enzyme-Linked ImmunoSorbent Assay) method. In the present work we propose a new method for assessing erythrocyte membranes using the SPR without the application of specific proteins.

2.2 Description of the SPR experiment

If a semitransparent Au film is placed at the interface (Fig.1), then under conditions of total internal reflection the SPR can occur. In this case, a polarized light beam is reflected on the interface of a noble metal surface

* Corresponding author: o.shynkarenko@isp.kiev.ua

DOI: 10.1051/epjconf/201713305002

vicinal dielectric layer at an angle superior to the total internal reflection angle.

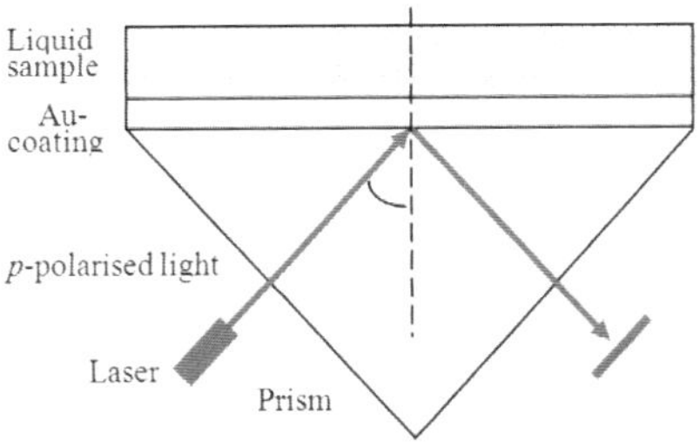

Fig. 1. Kretchmann prism coupling configuration for the SPR exitation.

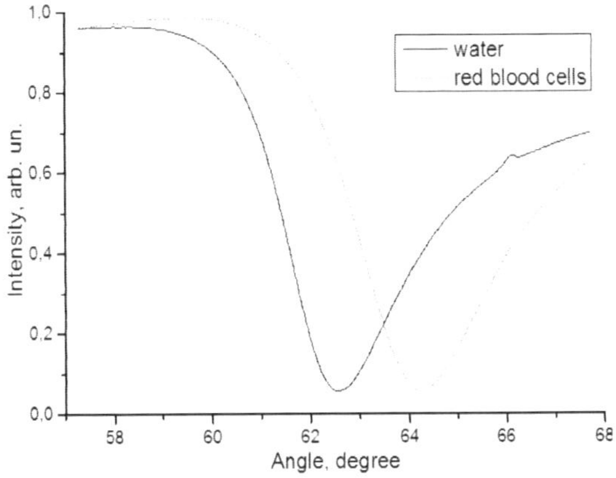

Fig. 2. Intensity of the reflected light versus angle of incidence for water and erythrocyte mass.

The underlying principle behind SPR is that a thin semitransparent layer of gold (or silver) will act as a mirror and reflect monochromatic p-polarised light projected at it except under special "matching" conditions. At the resonant angle the light energy will not be reflected, but rather converted into a collective oscillation of electrons in the metal film, generating the plasmon waveform. A direct plot of incident angle vs. reflected light intensity will show a marked loss of reflected energy at the resonant angle (Fig.2). The reflected light intensity will go sharply to almost zero at the resonant angle for a metal film of the right thickness.

2.3 Erythrocytes adsorption to the surface of the SPR sensor

The present paper devoted to study of the kinetics of erythrocytes adsorption to the surface of the SPR sensor. Resonance angle shift is proportional to the mass of the adsorbed film onto the surface of the sensor up to 200 nm in Z direction. Dimensions of erythrocytes are much more larger, approximately 2-7 microns, but under the influence of drugs affecting the stability of the membranes the erythrocytes change their properties that is displayed in the kinetic parameters of adsorption to the surface of the sensor. This can be conditioned by different types of packaging of erythrocytes near the surface as well as the erythrocytes membrane properties to change the shape or even just affect the kinetics of their interaction with the surface of the sensor.

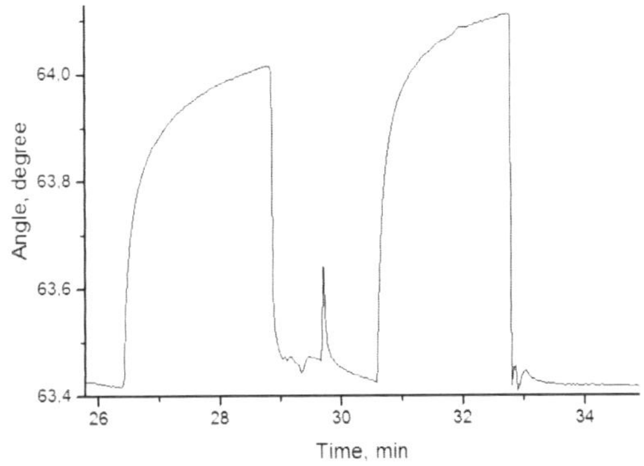

Fig. 3. Sensogram of SPR measurements for erythrocytes with addition of different CB.

An example of different SPR kinetics are given in Figure 3. The samples of erythrocyte mass of human blood are used in the investigation. This method requires a minimum pre-processing of fresh venous blood samples and explores the red blood cell mass obtained after centrifugation of samples.

3 Discussion and Conclusions

As is shown above obtained results for different gropes of patients proposed method enables us to registered an observed real deference in position of minimum of reflection in the incident angle scale. That was confirmed on the 50 number grope of patient at clinical researches.

We have also demonstrated that the addition of CB that affect the stability of membranes acts differently on the kinetics of erythrocytes ligament in the different groups of people. To compare the results of different patients the difference of signals of erythrocytes kinetics in blood samples of patients without adding special agents and with the addition of drugs that is used. In this case the impact of differences in the density of blood for different patients is excluded. This method is numerical, simple, reproducible and requires a minimum of time - 20-30 minutes for standardized methods, seemingly and it can be used to create a new methodology for determining the resistance of erythrocytes.

To determine the differences in the mechanism of interaction between erythrocytes under the influence of calcium channels the further research will be carried out.

The authors express their gratitude to employees of the Lviv State Cancer Center for providing blood samples for researches.

References

1. J. Suart, C. S. Johnson, *Baillibre's Clinical Haematology*, **l**. 747 (1987)

2. R.M. Hochmuth, R.E. Waugh, *Ann. Rev. Physiol.*, **49**, 209-19, (1987)

3. M.R. Clark, A.C. Greenquist and S.B. Shohet, *Blood,* **48**, 899 (1976)

4. C. Madigan and P. Malik, in book : Pathophysiology and Therapy for Haemoglobinopathies; Part I: Sickle Cell Disease, Vol. 8, (2006) pp. 1-23

5. T. Barnes, A. Shulman, A. Farone, M. Farone, D. Erenso, *Optics and Photonics Journal,* **3**, *211* (2013)

6. M. Dao, C.T. Lim, S. Suresh, *J. Mech. Phys. Solids*. **51,** 2259 (2003)

7. M.M. Brandao, A. Fontes, M. L. Barjas-Castro, L.C. Barbosa, F. F. Costa, C. L. Cesar, S. T. Saad, *Eur. J. Haematol*. **70,** 207 (2003)

8. W. Grünberg, J. A. Mol, E. Teske. J of Vet. Int. Med.**29**,395 (2015)

9. J. O. Oyewale, J of Vet. Med. A **41.1**, 62 (1994)

10. G. Del Mistro, S. Cervo, E. Mansutti, R. Spizzo, A. Colombatti, P. Belmonte, R Zucconelli, A. Steffan, V. Sergo, A. Bonifacio, *Analitic. and Bioanalitic. Chem.*, **407**, 3271 (2015)

11. M. Z. Vardaki, B. Gardner, N. Stone and P. Matousek,. *Analyst*, **140**, 5112 (2015)

AUTHOR INDEX